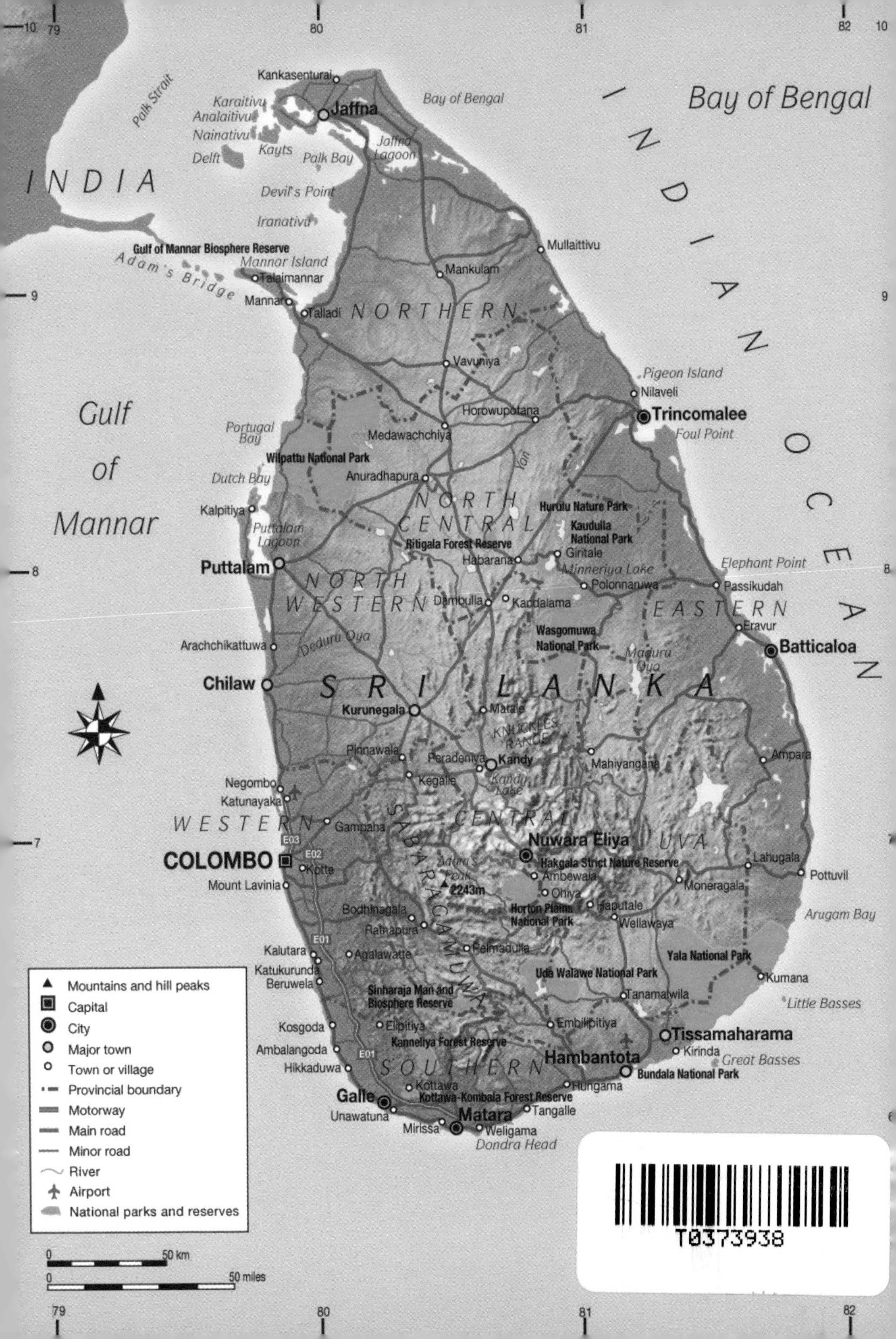
Bay of Bengal
INDIAN OCEAN
INDIA
Gulf of Mannar
Palk Strait
Palk Bay
SRI LANKA
NORTHERN
NORTH CENTRAL
NORTH WESTERN
EASTERN
CENTRAL
UVA
WESTERN
SABARAGAMUWA
SOUTHERN
Kankasenturai
Jaffna
Karaitivu
Analaitivu
Nainativu
Kayts
Delft
Jaffna Lagoon
Devil's Point
Iranativu
Gulf of Mannar Biosphere Reserve
Adam's Bridge
Mannar Island
Talaimannar
Mannar
Talladi
Mullaittivu
Mankulam
Vavuniya
Pigeon Island
Nilaveli
Trincomalee
Foul Point
Horowupotana
Medawachchiya
Portugal Bay
Wilpattu National Park
Dutch Bay
Anuradhapura
Kalpitiya
Puttalam Lagoon
Hurulu Nature Park
Kaudulla National Park
Ritigala Forest Reserve
Habarana
Giritale
Minneriya Lake
Puttalam
Elephant Point
Polonnaruwa
Passikudah
Dambulla
Kandalama
Eravur
Deduru Oya
Wasgomuwa National Park
Maduru Oya
Arachchikattuwa
Batticaloa
Chilaw
Kurunegala
Matale
Knuckles Range
Pinnawala
Peradeniya
Kandy
Kandy Lake
Mahiyangana
Ampara
Negombo
Kegalle
Katunayaka
Gampaha
Nuwara Eliya
COLOMBO
E03
E02
Kotte
Mount Lavinia
Adam's Peak
2243m
Hakgala Strict Nature Reserve
Ambewala
Ohiya
Lahugala
Pottuvil
Moneragala
Horton Plains National Park
Haputale
Bodhinagala
Ratnapura
Wellawaya
Arugam Bay
E01
Kalutara
Agalawatte
Pelmadulla
Yala National Park
Katukurunda
Beruwela
Uda Walawe National Park
Kumana
Sinharaja Man and Biosphere Reserve
Tanamalwila
Little Basses
Kosgoda
Elipitiya
Embilipitiya
Kanneliya Forest Reserve
Tissamaharama
Ambalangoda
Kirinda
Hambantota
Great Basses
Hikkaduwa
Bundala National Park
Hungama
Kottawa
Galle
Kottawa-Kombala Forest Reserve
Unawatuna
Matara
Tangalle
Mirissa
Weligama
Dondra Head
Mountains and hill peaks
Capital
City
Major town
Town or village
Provincial boundary
Motorway
Main road
Minor road
River
Airport
National parks and reserves
0
50 km
0
50 miles
79
80
81
82
10
9
8
7
T0373938

A PHOTOGRAPHIC FIELD GUIDE TO THE

AMPHIBIANS OF SRI LANKA

A PHOTOGRAPHIC FIELD GUIDE TO THE

AMPHIBIANS OF SRI LANKA

Anslem de Silva
Kanishka Ukuwela
Dilan Chathuranga

JOHN BEAUFOY PUBLISHING

First published in the United Kingdom in 2021 by John Beaufoy Publishing Ltd
11 Blenheim Court, 316 Woodstock Road, Oxford OX2 7NS, England
www.johnbeaufoy.com

10 9 8 7 6 5 4 3 2 1

Photo credits
Front cover: Leaf-nesting Shrub Frog © Anslem de Silva. **Back cover:** *top* Hollow-snouted Shrub Frog © Anslem de Silva; *bottom* Long-snout Tree-frog © Anslem de Silva. **Spine:** Common Indian Toad © Usui Toshikazu. **Title page:** Dull Green Shrub Frog © Usui Toshikazu.
Main descriptions: photos are denoted by a page number followed by a figure number..
Dilan Chathuranga 10(7); **Anslem de Silva** 9(1, 2), 10(4, 6, 8), 11(9), 13(11, 12, 13, 16, 17, 18), 15(20), 16(21), 24(24, 25), 25(26, 27, 28, 29), 28(30), 29(31, 32), 37(1, 2), 38(1), 40(1), 41(2), 43(3), 44(1), 45(2), 46(1), 47(3), 49(1, 3) 50(1), 51(2, 3), 52(1), 54(1), 55(2), 56(1), 57(2, 3), 60(1), 68(1),69(3), 72(1), 73(3), 74(1), 75(2), 76(1), 77(2, 3), 78(1), 79(2, 3), 81(2, 3), 83(2, 3), 84(1), 85(3), 86(1), 87(3), 89(2, 3), 91(2, 3), 95(2), 98(1), 99(3), 101(1), 103(2, 3), 107(2), 109(2, 3), 111(2, 3), 118(1), 123(2), 130 (1), 131(3), 133(2, 3), 135(3), 137(2, 3), 139(2, 3), 140(1),141(2, 3), 146(1), 160(1), 163(2, 3), 174(1), 175(2, 3), 179(3), 184(1), 185(2, 3), 187(1, 3), 188(1), 195(2, 3), 200(1), 201(3), 211(3), 218(1), 222(1), 223(2, 3), 224(1), 225(3), 227(3), 230(1), 231(2, 3), 233(2); **Suraj Goonewardene** 99(2), 134(1), 213(2, 3), 236(1); **Sameera Suranjan Karunarathna** 64(1), 65(2, 3); **Usui Toshikazu** 108(1), 122(1), 211(2); **Kanishka Ukuwela** 9(3), 10(5), 11(10), 13(14), 13(15), 17(22), 39(2, 3), 41(3), 42(1), 43(2), 45(3), 47(2), 49(2), 53(2, 3), 55(3), 58(1), 59(2, 3), 61(2, 3, 4), 62(1), 63(2, 3, 4), 66(1), 67(2, 3), 69(2), 71(1, 2, 3), 73(2), 75(3), 80(1), 82(2), 85(2), 87(2), 88(1), 91(1), 95(1), 97(1, 2), 99(2), 101(2), 102(1), 110(1), 113(1, 2), 117(1, 2), 119(2, 3), 123(3), 127(1, 2), 131(2), 132(1), 135(2), 136(1), 138(1), 143(1, 2), 144(1), 145(2, 3), 149(2), 158(1), 159(2, 3), 161(2, 3), 162(1), 164(1), 165(2), 171(1, 2), 173(1, 2), 174(1), 177(1, 2), 178(1), 179(2), 181(1, 2), 183(1, 2), 187(2), 189(2, 3), 193(1), 194(1), 196(1), 197(2, 3), 201(2), 203(1 & 2), 204(1), 205(2, 3), 207(1, 2), 210(1), 212(1), 216(1), 217(2, 3), 219(2, 3), 220(1), 221(2), 225(2), 226(1), 227(2), 229(1, 2), 233(1, 3), 237(2, 3); **L. J. Mendis Wickramasinghe** 65(4), 115(1, 2), 125(1, 2), 150(1), 151(2, 3), 153(1, 2), 169(1, 2), 191(1, 2), 209(1, 2); **Sanoj Wijayasekara** 93(1, 2), 104(1), 105(2), 106(1), 120(1), 121(2, 3), 128(1), 129(2), 147(2, 3), 148(1), 149(3),155(1, 2), 157(1, 2), 167(1, 2), 175(2, 3), 193(2), 198(1), 199(2, 3), 215(1, 2), 235(1, 2).

ISBN 978-1-913679-11-8

Edited by Krystyna Mayer
Designed by Gulmohur
Project management by Rosemary Wilkinson
Cartography by Krishantha Sameera De Zoysa
Illustrations by Eranga Geethanjana Perera (EGP)

Printed and bound in Malaysia by Times Offset (M) Sdn. Bhd.

Contents

Preface

During the past two decades, interest in herpetology and especially in amphibians of Sri Lanka has increased, and as a result many people are currently involved in active research on their different aspects. However, a well-illustrated guide to amphibians of the country has been long overdue. The last treatise on the subject in the English language was published more than 25 years ago by Sushil Dutta and Kelum Manamendra-Arachchi, while a much more updated version was compiled by Kelum Manamendra-Arachchi and Rohan Pethiyagoda and published in the Sinhala language nearly 15 years ago. Anslem de Silva published a *A Photographic Guide to Common Frogs, Toads and Caecilians* in 2009. Since these publications about 15 species have been described, and numerous changes in taxonomy have taken place. The immediate need for an updated identification guide to the amphibians of Sri Lanka was apparent – and was stressed by Professor Priyanka de Silva of the University of Peradeniya, who wanted a simple updated identification guide to help teach undergraduate and postgraduate students in zoology. We are sincerely thankful to her for the suggestion, which led to this title.

This book features species that may be encountered in both anthropogenic and pristine habitats. Species currently considered to be extinct and very rare are also listed for completeness. It must be stressed that this book is not a comprehensive guide for specialist readers in the field. A specialist who needs to confirm identities of 'taxonomically difficult species' should further refer to the original research papers that dealt with the subject. Nevertheless, it is hoped that this book will instill interest in amphibians in future generations of herpetologists in Sri Lanka, and inspire further studies on the taxonomy, ecology and conservation of these delightful creatures.

Anslem de Silva
Kanishka Ukuwela
Dilan Chathuranga
30 November 2020

Foreword

The need for a guide to the amphibian fauna of Sri Lanka is obvious, given the currently critical condition of these organisms. Amphibians are an attractive group of animals whose diversity has always sparked interest among the scientific community, creating a vast collection of unanswered questions. However, the identification of amphibians has been a challenge due to the lack of a complete and informative guide. This comprehensive pictorial guide should thus be of great benefit to a better understanding of the unique and intriguing nature of these fascinating animals.

Much-valued commendation and admiration go out to the authors, who have done an outstanding job in compiling this book. An introduction to the guide briefly describes the history, current status, threats and conservation information, along with interesting folklore associated with amphibians. With the clear and informative images, distribution maps and updated status of each species, this guide can easily be comprehended by experts and beginners in the field alike.

I firmly believe that this book will be very useful to undergraduate and postgraduate students in the field of zoology, biology and environmental science, as well as researchers, wildlife managers and visitors.

W.A. Priyanka P. De Silva, PhD (USA)
Professor in Zoology, Faculty of Science
University of Peradeniya, Sri Lanka
30 November 2020

About This Book

Like our previous guide to the reptiles of Sri Lanka, *A Naturalist's Guide to the Reptiles of Sri Lanka* (de Silva & Ukuwela, 2017, 2020), this book is intended for both naturalists and visitors to Sri Lanka, providing an introduction to the amphibians of Sri Lanka. It features all the extant species of amphibian in Sri Lanka, with colour photographs and quick and easy tips for identification. At the time of writing, 120 species have been recorded in Sri Lanka, and ongoing taxonomic work is certain to add more to this impressive list within the next few years.

This guide provides a general introduction to the amphibians of Sri Lanka, a profile of the physiographic, climatic, and vegetation features of the island, key characteristics that can be used in the identification of amphibians and descriptions of each extant amphibian species. Additionally, it presents information on amphibian conservation in Sri Lanka and a brief introduction to folklore and traditional treatment methods for poisoning due to amphibians in Sri Lanka. The species descriptions are arranged under their higher taxonomic groups (orders and families), and further grouped in their respective genera. The descriptions are organized in alphabetical order by their scientific names. Every species covered is accompanied by one or more colour photograph of the animal. Each account includes the vernacular name in English, the current scientific name, the vernacular name in Sinhala, a brief history of the species, a description with identification features, and details of habitat, habits and distribution (in Sri Lanka and outside the country).

Key external identification features of the species, such as body form, skin texture and colouration, are provided, to help in the quick identification of an animal in the field. However, for some species, like the shrub frogs of the genus *Pseudophilautus*, features such as the presence of a lingual papilla, and supernumerary tubercles on the palms and feet, are also provided. Although these characteristics are difficult to see in the field, this information has been included since in some instances it is needed to confirm the identity of a species. However, it must be noted that according to Sri Lanka's wildlife laws, amphibians cannot be captured or removed from their natural habitats without official permits, which must be obtained in advance from the Department of Wildlife Conservation.

Note that this work is not comprehensive, and more specialist readers should confirm details of identification with more technical works on the subject (references, p. 241). The websites 'Amphibian Species of the World', 'AmphibiaWeb' and 'Amphibians of Sri Lanka' can also be consulted for recent revisions in the taxonomy of amphibians. The authors may be contacted via email for further information.

Anslem de Silva: kalds@sltnet.lk
Kanishka Ukuwela: kanishkauku@gmail.com
Dilan Chathuranga: dilanchathuranga9@gmail.com

General Introduction to Amphibians

Amphibians are ectothermic tetrapod vertebrates with moist or dry glandular skin. They belong to the vertebrate class Amphibia and are distributed all over the world except Antarctica. All amphibians pass through a larval stage before developing (metamorphosing) into adults. The larval stage can be aquatic or can sometimes be completed totally within the egg (direct development), which is usually laid on land. Currently (2020), there are more than 8,150 species of amphibian in the world, but the number keeps growing at a steady rate as many new species are discovered, especially in the tropics. Amphibians that are living today are called modern amphibians and are grouped in the amphibian subclass Lissamphibia. The lissamphibians are grouped into three orders: Anura, Caudata (Urodela) and Gymnophiona (Apoda). The order Anura comprises frogs, while Caudata consists of newts and salamanders, and Gymnophiona of limbless amphibians, termed caecilians.

Fig. 1 *A calling male frog*

Compared to other vertebrates (mammals, fish, reptiles and birds), amphibians have many general features and habits unique to them. Some are attractively coloured (bright green, red and so on), and they can rapidly change colour according to the surrounding substrate, environment or even time of day. The majority of amphibians have a sticky tongue that is located at the front of the lower jaw, which can be flipped out to capture prey.

Frogs, toads and caecilians have diverse and interesting methods of reproduction. There are at least 36 modes of reproduction known among amphibians. Fertilization is external in the vast majority of amphibians, except in the Caudata and Gymnophiona. Breeding activity in frogs usually begins with males calling (Fig. 1) either singly or in groups at breeding sites.

While each species has its unique call (which is currently used as one taxonomic criterion to identify species), some do not call at all. A mature female will respond to the call and approach the male, which is usually smaller than the female. The male mounts her and

Fig. 2 *Axillary amplexus in a pair of frogs*

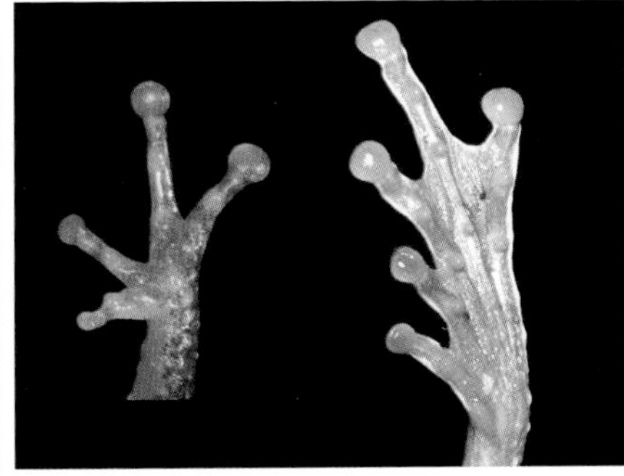

Fig. 3 *The expanded digit-tips, 'discs' of a tree frog*

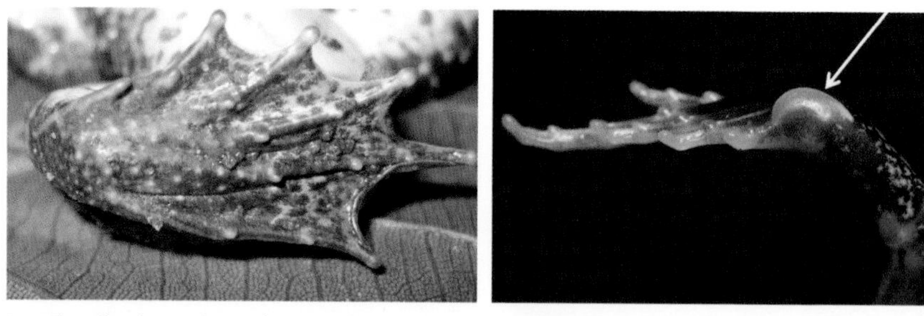

Fig. 4 The webbing between the toes of an aquatic frog

Fig. 5 Shovel-shaped tubercle (inner meta-tarsal tubercle)

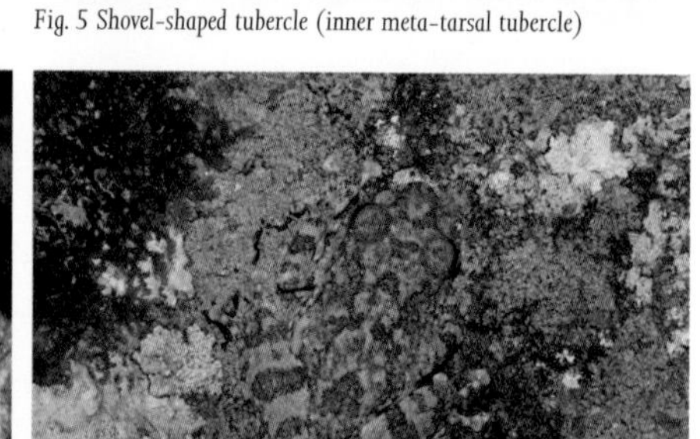

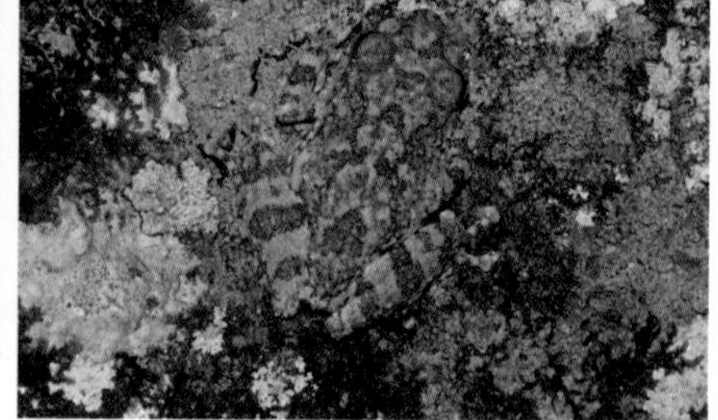

Fig. 6 Parotid glands at the back of the head of a toad

Fig. 7 A shrub frog camouflaged with its moss-covered tree trunk

clasps her tightly with his forelimbs, which is known as amplexus. Most males usually have a sticky pad on the base of the thumb (nuptial pad), which helps them to get a firm grip on the female. Amplexus is carried out in many ways, and axillary amplexus, where the male holds or grips the female just behind the forelimbs (Fig. 2) is one of the common types. When the female begins laying eggs, the male simultaneously releases sperm to fertilize them.

Frogs have long, muscular hindlimbs that power their main mode of locomotion, jumping. Toads usually do not jump, but instead take short hops or even 'walk'. Frogs' toes are good indicators of their main habitat – for example, the tips of the fingers and toes of tree and shrub frogs are expanded into pads or circular discs (Fig. 3) that help them to cling to the surface. Aquatic species have well-developed webbing between the toes (Fig. 4) to propel themselves in water, and fossorial species have special 'shovel-like' modified appendages on the feet (Fig. 5) that help them to dig or burrow into the soil.

Frogs and toads (including tadpoles) have various interesting strategies to defend themselves from predators like

Fig. 8. A toad inflating its body to ward off predators

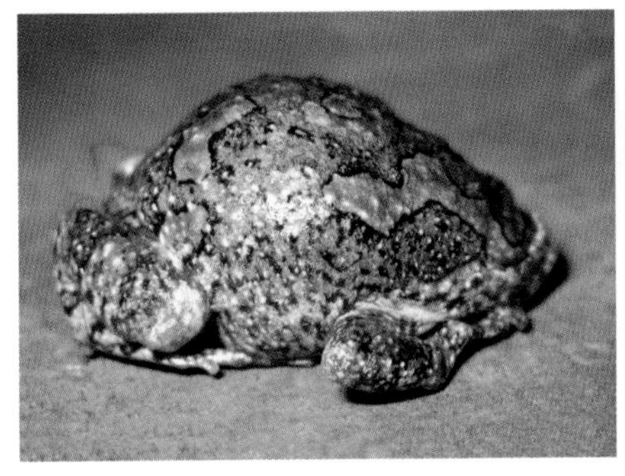

Fig. 9 *Uperodon taprobanicus* with head sunk into lower jaw

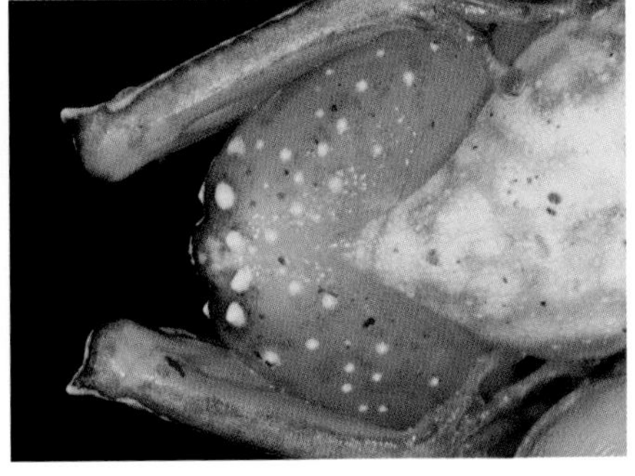

Fig. 10 Spiny tubercles around the cloaca and spur-like calcars on heels

birds, mammals, snakes and possibly other large frogs. The most common method used by frogs is to leap away from danger. Some species even squirt a jet of 'urine' on the predator during the leap. Some frogs have toxins in their skin glands that make them poisonous to their predators. In toads, the toxin glands are localized into regions on top of the neck called parotid glands (Fig. 6).

Some frogs (like narrow-mouth frogs) exude toxic, sticky and smelly secretions from the surface of the body. Another important and widely used method of defence involves blending into their surroundings (camouflage) (Fig. 7). Some species inflate the body, appearing to be several times larger than their normal size (like *Uperodon systoma*, *U. taprobanicus* and species of Sphaerotheca) (Fig. 8). Some frogs (such as *Uperodon taprobanicus*) virtually sink the head into their lower jaw, and hide the head and face (Fig. 9). The body structures of certain frogs deceive their predators. Tree-frogs of the genus *Taruga* have white spiny tubercles around the cloaca that appear like teeth in a mouth, and spur-like calcars on the heels resembling eyes (Fig. 10). In *Microhyla mihintalei* there are dark brown spots that resemble a 'face' from the rear side of the frog. This type of colouration might startle a predator that may be approaching from the back. Some species feign death when they are threatened to discourage predators from eating them, as predators usually prefer fresh prey.

Introduction to Sri Lanka's Amphibians

Sri Lanka is a global amphibian hotspot that currently boasts approximately 120 species. However, this number keeps growing as many new species are being discovered on a yearly basis. Ongoing studies by us and work in progress by many other researchers indicate that there are yet more amphibian species to be scientifically described in Sri Lanka. However, for an island of 65,610km^2 in area, the species diversity is remarkable.

Of the 120 amphibian species in Sri Lanka, 107 (89 per cent) are endemic. Sri Lankan amphibians are represented only by the amphibian orders Anura and Gymnophiona. The order Anura in Sri Lanka is represented by the six frog families Bufonidae, Dicroglossidae, Microhylidae, Nyctibatrachidae, Ranidae and Rhacophoridae. The order Gymnophiona, which includes caecilians, is represented by a single family, Ichthyophiidae. Interestingly, the Anuran genera, *Adenomus*, *Lankanectes* and *Nannophrys*, are endemic to the country (Table 1).

Molecular phylogenetic studies have shown that Sri Lanka has maintained an amphibian fauna that is distinct from that of the Western Ghats in India, and some species (genus *Lankanectes*) even have ancient origins dating back to the Cretaceous period. The genus *Pseudophilautus* also consists of a large endemic insular radiation that has evolved in the island.

Table 1. Summary of Amphibian Diversity in Sri Lanka
(As of December 2020)

Order	Family	No. Species	Genus	No. Species	Endemic species
Anura	Bufonidae	6	*Adenomus*	2	2
			Duttaphrynus	4	2
	Dicroglossidae	13	*Euphlyctis*	2	0
			Minervarya	3	2
			Hoplobatrachus	2	0
			Nannophrys	4	4
			Sphaerotheca	2	0
	Microhylidae	10	*Microhyla*	4	3
			Uperodon	6	4
	Microhylidae	2	*Lankanectes*	2	2
	Ranidae	3	*Hydrophylax*	1	1
			Indosylvirana	2	2
	Rhacophoridae	83	*Pseudophilautus*	77	77
			Polypedates	3	2
			Taruga	3	3
Gymnophiona	Ichthyophiidae	3	*Ichthyophis*	3	3
Total		120		120	107 (89%)

General Habits

Based on where they live, amphibians of Sri Lanka can be grouped as aquatic (for example Ranidae and Dicroglossidae), terrestrial (Bufonidae and others), fossorial (like Ichthyophiidae and Microhylidae) and arboreal (Rhacophoridae). Nevertheless, the terrestrial forms may also exhibit fossorial or aquatic tendencies, and vice versa. Most endemic species belonging to these four groups are found in forests of the wet zone, including lowland rainforests and montane forests.

Sri Lankan amphibians show a great diversity of reproductive strategies. This is even evident starting from where they lay eggs. Some lay eggs in stagnant water embedded in a separate gelatinous mass and the egg clusters float on the surface for about 24–35 hours (Fig. 11). Some lay eggs in long strands in water, which get entangled with the aquatic vegetation (toads, for example) (Fig. 12). Some species of the genus *Polypedates* and *Taruga* deposit eggs in foam nests (Fig. 13). These are created by beating the fluids secreted by the female to a lather that is then suspended above the water's surface on the ground, on vegetation or on man-made structures like water tanks, close to water. The tadpoles emerging from the eggs wriggle out of the foam nest and drop into the water below. Some microhylids deposit their eggs a few centimetres above the water level inside tree-holes (for example *Uperodon nagaoi*) (Fig. 16). Similarly, frogs of the genus *Nannophrys* lay eggs on moist rock surfaces.

All bufonids, ranids, dicroglossids, microhylids and some rhacophorids have an aquatic larval stage known as tadpoles (Fig. 14). Tadpoles are surface, mid-water or bottom

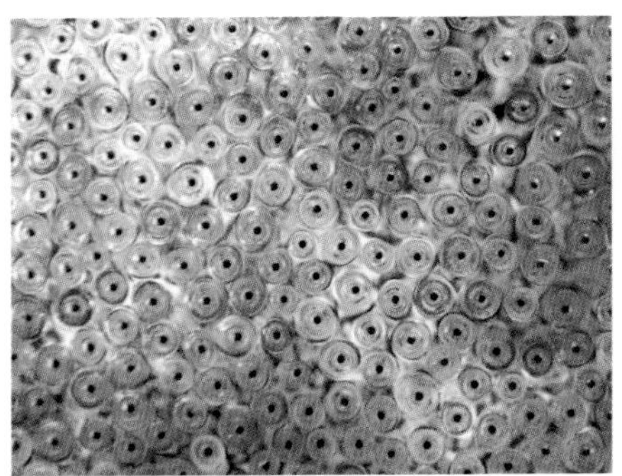

Fig. 11 *A gelatinous mass with egg clusters*

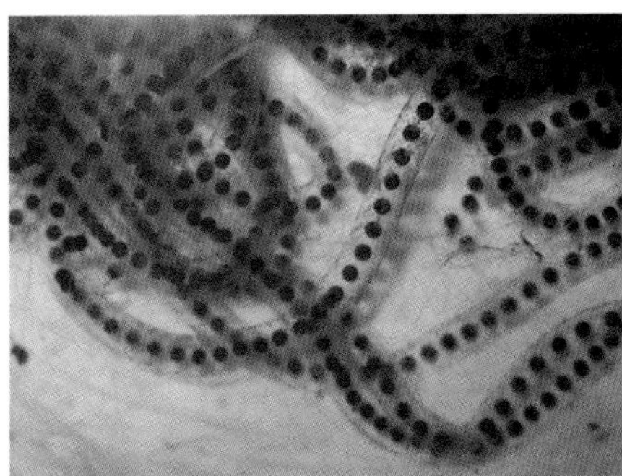

Fig. 12 *Long strands toad spawn in water*

Fig. 13 *A foam nest of a* Polypedates *species*

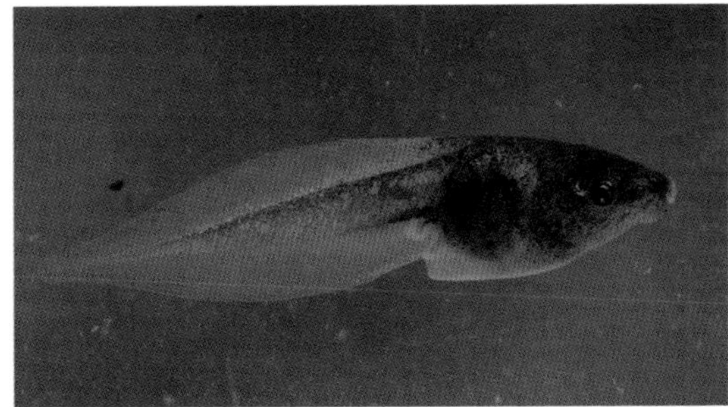

Fig. 14. *A tadpole*

Fig. 15 *Direct-developing egg mass buried in soil*

Fig. 16 *Eggs of* Uperodon nagaoi *deposited above water in a tree-hole*

Fig. 17 *A female* Ichthyophis glutinosus *with freshly laid eggs*

Fig. 18 *Developing larvae of* Ichthyophis glutinosus *within the eggs*

feeders that feed on vegetation, detritus or aquatic macroinvertebrates (hence they are known as exotrophic tadpoles). However, all known species of the genus *Pseudophilautus* are direct developing (that is, they undergo the tadpole stage within the egg, emerging as metamorphosed imagos or diminutive frogs). Nesting in direct developers takes place in a hole dug in humus (Fig. 15), under leaf litter on the forest floor, or on the underside of foliage (for example *Pseudophilautus femoralis*). The nutrition for the development and metamorphosis is provided by the egg yolk, hence the tadpoles are known as endotrophic tadpoles. Tadpoles of the rock frogs (*Nannophrys*) live on wet rock surfaces. To enable this, they hatch at a later stage of development where they have already developed hindlimbs and thus have the ability to crawl on the wet rock surfaces they live on.

Caecilians lay their eggs inside earth cavities, or under stones or decaying logs, and the female usually remains with the eggs, most probably to 'guard' them (Figs 17–18). Similarly, the male of the rock frog *Nannophrys ceylonensis* is known to stay close to the eggs, probably to guard them. It is also speculated that the extinct shrub frog *Pseudophilautus maia* carried the direct-developing eggs attached to its belly.

Geoclimatic Profile of Sri Lanka

Knowledge and understanding of the physiography, climatic zones, vegetation types and aquatic resources of Sri Lanka are important for the study of amphibians, as these

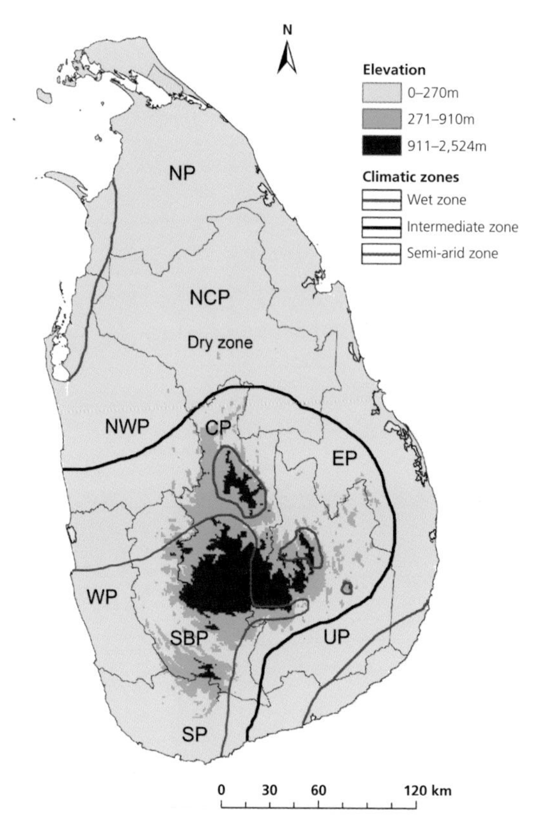

Fig. 19 Physiography, climatic zones and administrative provinces of Sri Lanka

parameters greatly influence their distribution and activity. This information also assists in the understanding of the habitat requirements of particular amphibian species.

Sri Lanka is a humid tropical island in the South Asian region, situated in the Indian Ocean, off the southern tip of the Indian peninsular at latitudes 5°55'–9°51' N and longitudes 79°41'–81°54' E. The island is 65,610km^2 in area, of which 64,740km^2 is land and the rest is inland waters.

Physiography

Sri Lanka consists of three distinct peneplains, or erosion levels, recognized according to elevation and slope features. The lowest or first peneplain is the largest, and extends inland from the coast to an elevation of 270m above mean sea level. The second peneplain, or the uplands, extends from 270m to about 910m, and occupies nearly three-tenths of the island. The highlands or the third peneplain lie at elevations of 910–2,524m, centred towards the south centre of the island (Fig. 19).

Climatic Zones

There are four distinct climatic zones on the island, which are mainly designated by the annual rainfall received. These are the semi-arid zone, dry zone, intermediate zone and wet zone (Fig. 19). The **dry zone** constitutes 55 per cent of the island's total land area, with a rainfall of 1,250–1,900mm a year and temperature of 27–30° C. The **wet zone** constitutes 23 per cent of the island's total land area, with a rainfall of 2,500–5,000mm a year. The humidity in this region is 75–85 per cent. The **intermediate zone** constitutes 12 per cent of the island's total land area and lies between the dry zone and the wet zone. The rainfall is 1,900–2,500mm a year. It is considered as a transition zone between the wet and dry zones. The **semi-arid zone** receives about 1,000–1,250mm of rainfall annually and comprises nearly 10 per cent of the island's total land area.

Vegetation & Natural Forest Cover

Sri Lanka has a rich flora with nearly 3,150 species of flowering plant, of which 25 per cent are endemic. The island's vegetation is divided into the following types: tropical dry

Fig. 20 Tropical lowland wet evergreen rainforests of Sinharaja

mixed evergreen forests, tropical lowland wet evergreen rainforests, tropical submontane wet evergreen rainforests, tropical montane wet evergreen rainforests, tropical moist semi-evergreen forests, tropical thorn forests, savannah forests, grassland and mangroves. The natural forest cover of Sri Lanka in 1881 was about 84 per cent of the island. However, due to the clearing of natural forests for agriculture (coffee, tea, rubber, coconut and so on), timber, and other human activities, the natural forest cover of the island had declined to 23.8 per cent by 1992 and to 17 per cent by 2018. As a result, only a small range of the former extent of forests is left. Characteristics of some of the major vegetation types of Sri Lanka are described below.

- **Tropical dry mixed evergreen forests** are found in the lowland (elevation 0–910m above sea level) dry zone of Sri Lanka. These forests are dominated by species of *Manilkara, Chloroxylon, Vitex, Berrya* and *Schleichera*. Due to the presence of some deciduous species, these forests are called semi-evergreen forests or monsoonal forests. It is believed that dry mixed evergreen forests developed in the dry zone over the past 500–800 years. Due to the dry conditions in these forests, the diversity of amphibians is low and their abundance depends on seasonality.

- **Tropical lowland wet evergreen rainforests** are distributed in the lowland wet zone to an elevation of up to 900m above sea level. These forests are dominated by trees of the Dipterocarpaceae family and are found in the southwestern region of Sri Lanka. Tall trees with an emergent layer and well-defined canopy are characteristic features of these forests. They are home to a high diversity of endemic amphibians. Sinharaja, Kanneliya, Kithulgala, Dombagaskanda and Kottawa contain well-known examples of these types of forest. The best-known and largest remaining tract of lowland wet evergreen rainforest in Sri Lanka is the Sinharaja rainforest. This unique forest was added to the World Heritage list in 1989 (Fig. 20).

Fig. 21 *An inland lake in the dry zone*

Fig. 22 Tropical montane wet evergreen rainforests

- **Tropical submontane wet evergreen rainforests** occur in mid-elevations of the wet zone (900–1,500m above sea level) of the Central Hills of Sri Lanka. They are characterized by relatively shorter trees with a dense canopy, with the dominant trees being *Shorea, Syzgium, Calophyllum, Cullenia* and *Myristica*. The mid-elevations of Sri Pada Sanctuary and Knuckles Mountain Range and Loolecondera forest contain a few forests of this type. They are also areas with a high diversity of endemic amphibians, especially for the genus *Pseudophilautus*.

- **Tropical montane wet evergreen rainforests** are distributed at 1,500–2,500m above sea level in the Central Hills in the wet zone of Sri Lanka. They are dominated by tree species of the genera *Calophyllum, Syzgium, Eugenia, Litsea* and *Myristica*. The upper regions of Sri Pada Sanctuary and Knuckles Mountain Range (Fig. 22), Horton Plains, Hakgala Strict Nature Reserve and Piduruthalagala contain some of the well-known forests of this type. They are home to a high diversity of *Pseudophilautus* species.

Aquatic Habitats

Sri Lanka possesses 3ha of inland lentic waters for every square kilometre of land. This is considered one of the highest densities of inland waters in the world. There are about 12,000 man-made irrigation lakes (tanks) located in the dry zone (Fig. 21). The first lakes were constructed in around the sixth and fifth centuries BC. They cover an area of approximately 170,000ha. There are nine major rivers and 94 small rivers (their collective length is about 4,560km) that drain into the Indian Ocean. These irrigation lakes, rivers and their tributaries are ideal habitats for aquatic frogs.

Identification of Sri Lanka's Amphibians

This section describes the key diagnostic features of amphibians that are used in the species descriptions. It also includes an easy-to-use illustrated identification guide to Sri Lanka's amphibian genera. This information helps in the identification of an amphibian to the generic level in the field.

Most people can readily differentiate a frog from a reptile. However, many may find it difficult to believe that a snake-like or eel-like caecilian is in fact an amphibian. Caecilians can be easily distinguished from snakes by the lack of scales on the body in caecilians, and from eels by the lack of fins in caecilians.

Key to Main Diagnostic Features

AST Antepenultimate Subarticular Tubercle; **CAL** Calcar; **CAS** Canthal Ridge; **CUF** Cutaneous Fringe; **DEF** Dermal Fringe; **DGF** Dorsolateral Glandular Fold; **DIS** Disc; **DST** Distal Subarticular Tubercle; **EYE** Eye; **FEM** Femur; **FI1–FI4** Fingers 1–4; **FLA** Flank; **GUS** Gular Sac; **IDR** Interupted Dorsal/Longitudinal Ridge; **IMT** Inner Metatarsal Tubercle; **INR** Internarial

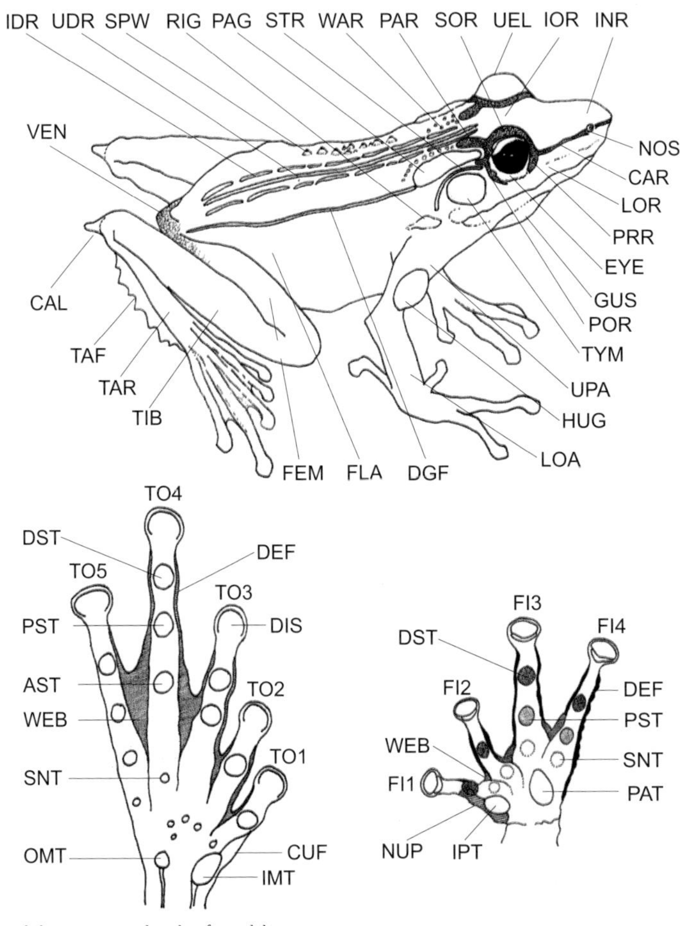

Fig. 23 External characteristics used to identify amphibians

Region; **IOR** Inter-Orbital Region; **IPT** Inner Palmer Tubercle; **LOA** Lower Arm; **LOR** Loreal Region; **NOS** Nostril; **NUP** Nubtial Pad; **OMT** Outer Metatarsal Tubercle; **PAG** Parotid Glands; **PAR** Parietal Ridge; **PAT** Palmer Tubercle; **POR** Postorbital Region; **PRR** Preorbital Ridge; **PST** Penultimate Subarticular Tubercle; **RIG** Rictal Glands; **SNT** Supernumerary Tubercle; **SOR** Supraocular Ridge; **SPW** Spiny Warts; **STR** Supratympanic Ridge; **TAF** Tarsal Fold; **TAR** Tarsus; **TIB** Tibia; **TO1–TO5** Toes 1–5; **TYM** Tympanum; **UDR** Uninterupted Dorsal/Longitudinal Ridge; **UEL** Upper Eyelid; **UPA** Upper Arm; **VEN** Vent; **WAR** Warts; **WEB** Webbing.

Table 2. Guide to Sri Lanka's Amphibian Genera

Genus	External Features	
Adenomus	Skin moist and granular with numerous warts. Elongated parotid glands behind the eyes. Body slender and flat. Digits end without discs. Finger and toe margins smooth.	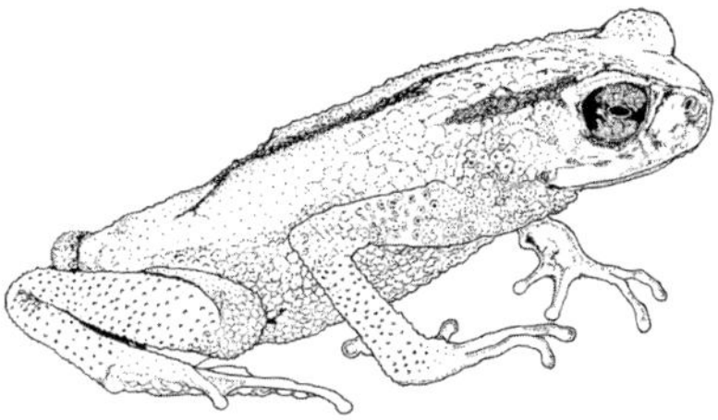
Duttaphrynus	Skin dry with numerous warts. Body stout. Parotid glands behind eyes. Limbs short and digits end without discs. Finger and toe margins serrated.	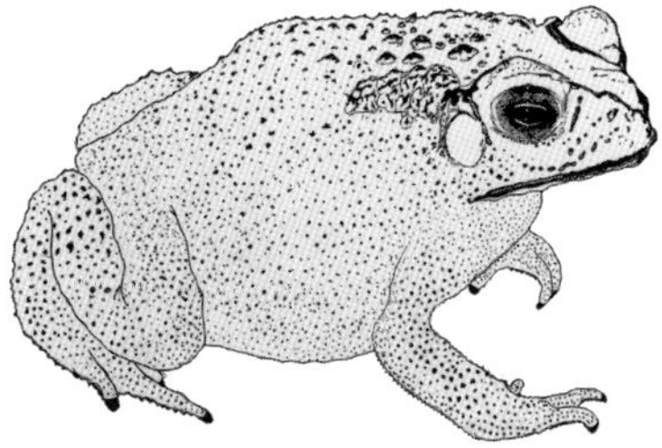
Microhyla	Body small (SVL <35mm) and triangular. Skin moist with numerous small tubercles. Narrow head region with small mouth. Digits without prominent discs.	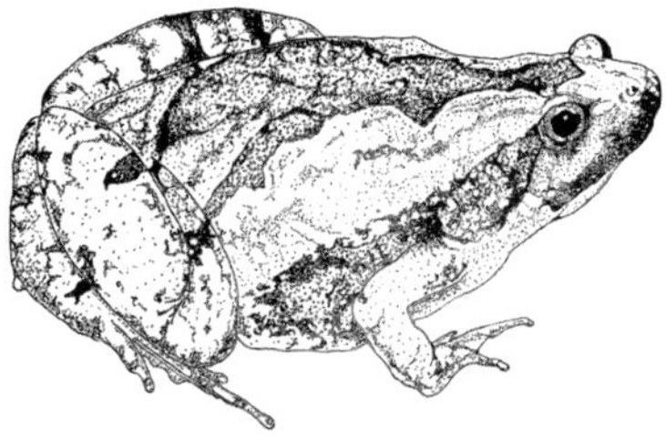

Uperodon	Body small (SVL <50mm) to stout to flat elongated, balloon-like when bloated. Narrow head region with small mouth. Skin moist with numerous small tubercles. Fingertips have triangular discs (except in *U. systoma*).	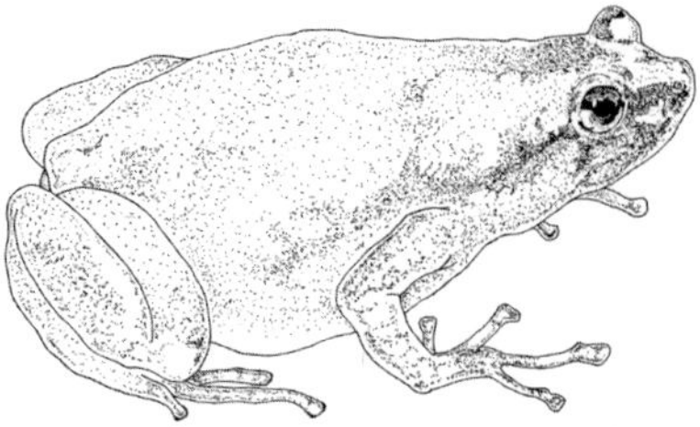
Euphlyctis	Body moderate to robust (SVL <90mm) and elongated. Skin smooth, and eyes bulging and pointing upwards. Toes fully webbed.	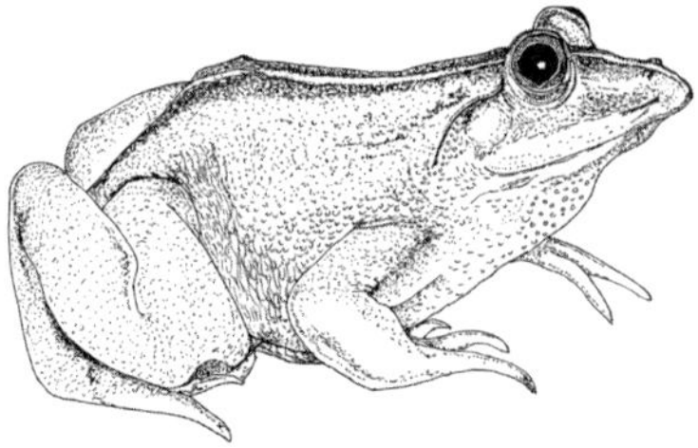
Hoplobatrachus	Body large (SVL >60mm), Dorsum has longitudinal ridges. Toes fully webbed.	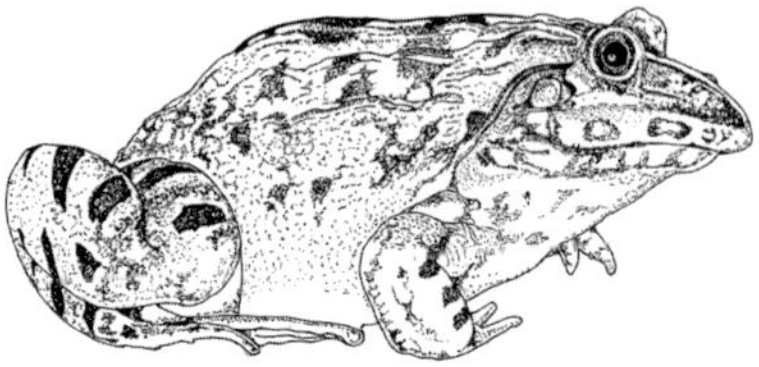

Minervarya	Body small to moderate (SVL 20-40mm) and elongated. Dorsal skin with longitudinal folds or warts in lines. Toes half to two-thirds webbed.	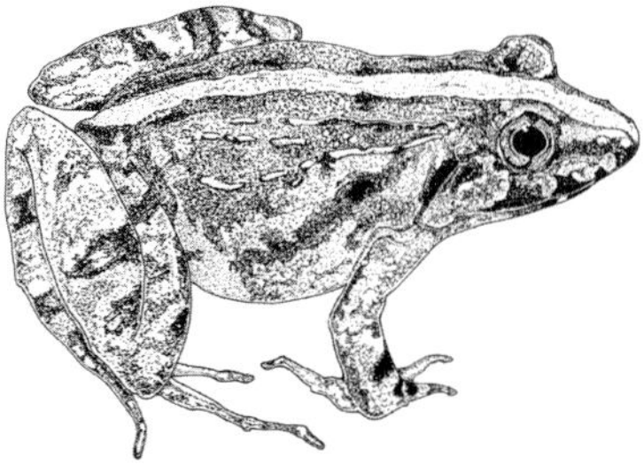
Sphaerotheca	Body stocky, toad-like but with moist skin. Snout blunt. Special 'shovel-like' modified tubercle/appendage on heel. Toes half to two-thirds webbed.	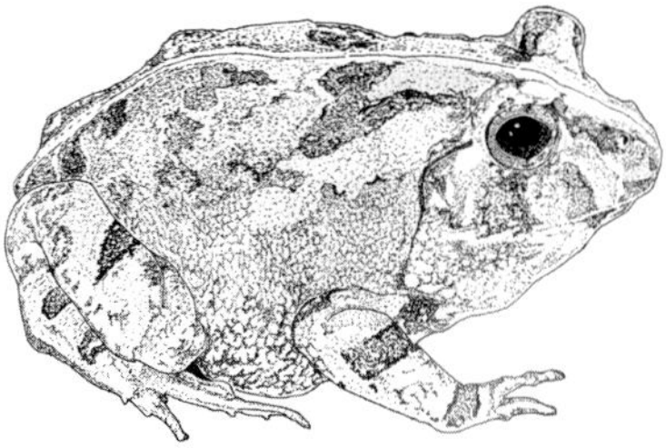
Nannophrys	Body small to moderate and depressed. Head broad and flat. Skin with small nodules. Digits end without discs. Toes half to two-thirds webbed.	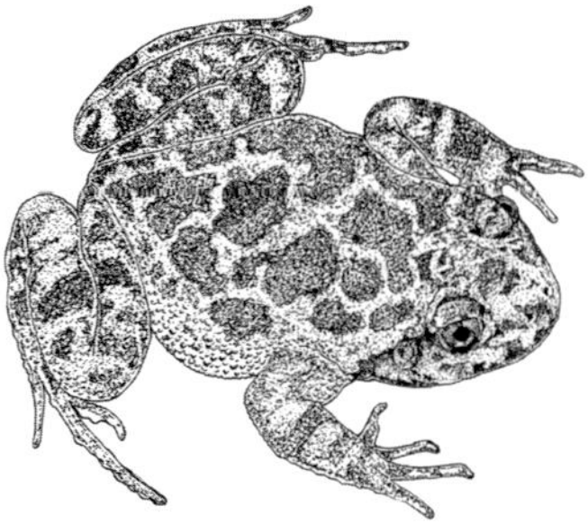

Hydrophylax	Body moderate and elongated. Distinct dorsolateral glandular fold extends from behind eye to vent. Thick glandular white to gold line from back of mouth to flank. Longitudinal stripe on shank. Toes half to two-thirds webbed. Tips of digits slightly enlarged.	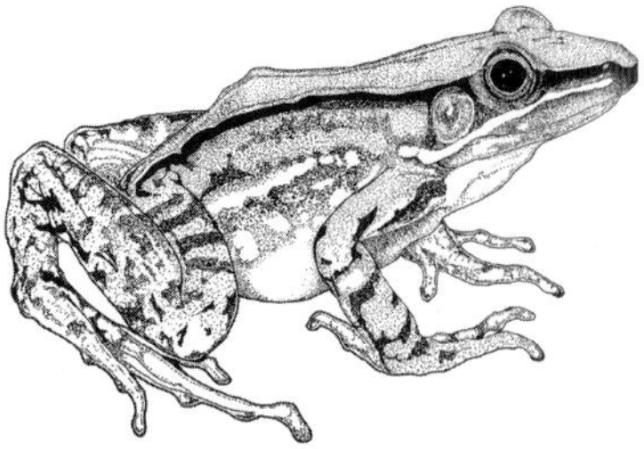
Indosylvirana	Body moderate to large. Distinct dorsolateral glandular fold extends from behind eye to region of vent. Toes half to two-thirds webbed. Tips of digits slightly enlarged.	
Polypedates	Body moderate (SVL <80mm) and elongated. Long hindlimbs. Toes medially webbed, and digits end in distinct circular discs. Large eyes, horizontal pupils.	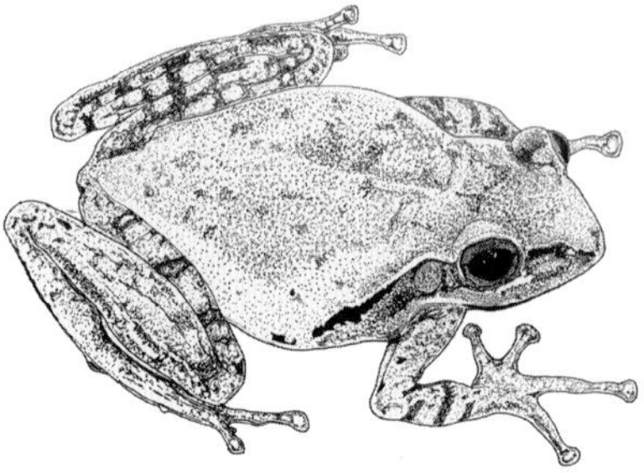
Pseudophilautus	Body small to moderate (SVL <50mm). Highly variable features. Long hindlimbs. Toes a quarter to two-thirds webbed. Digits end with discs. Large eyes, horizontal pupils.	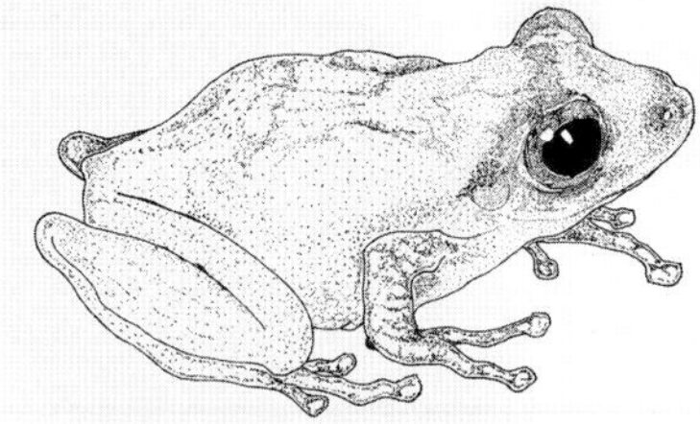

Taruga	Body moderate and elongated. Toes medially webbed, and fingertips with distinct circular discs. Head elongated and snout pointed. Hindlimbs long with short, distinct spine (calcar) at heel. Prominent dorsolateral fold from back of eye to mid-flank. Conical tubercles around cloaca.	
Ichthyophis	Long, limbless, cylindrical, scale-less slimy body with distinct grooves throughout. Eyes present, but highly reduced. Distinct head and mouth. Small eversible tentacles between eyes and nostrils.	

Conservation

Amphibians all over the world are facing an extinction crisis, and are considered to have the highest extinction rate of any vertebrate group. According to some estimates, almost a third of the world's amphibians are threatened with extinction. Numerous potential causes for these threats have been proposed, such as habitat degradation, fragmentation and loss, introduced species, pollution, pathogens and diseases, and climate change. Most of these threats are due to human activity, which is generally believed to exacerbate extinction risks. Although it is thought that 35 of the world's amphibian species have become extinct since the 1500s, this number is now considered by the International Union for the Conservation of Nature (IUCN) to be closer to 130, with 122 species having disappeared since the 1980s. Amphibians are at increased risk of the effects of environmental stresses because of their limited home ranges, their trophic position and potential for bioaccumulation, and the ease with which chemicals are transported across their skin throughout various stages of development. During the past two decades, the rapid decline of amphibian populations has caused global concern. Amphibians are considered to be one of the best indicators of environmental degradation. Thus, the decline in amphibian populations the world over indicates deterioration of the global environment.

Of the 130 amphibian extinctions known to have occurred across the world, sadly 18 (14 per cent) have occurred in Sri Lanka. This is one of the highest number of amphibian extinctions known from a single country. Some consider this unusual extinction rate to be largely the result of the loss of approximately 70 per cent of the island's forests. This assumption fits in well as wide tracts of virgin forests (montane, submontane, and lowland and other forests) were cleared for coffee, cinchona, tea, rubber plantations and so on by British colonists commencing in the early nineteenth century. At the time, specimens collected in Sri Lanka were sent to various European museums (especially the British Museum). Recently, when some of the specimens in European museums that were collected in 1850–1940 were re-examined, they turned out to be new species that did not have a name. Sadly, when they were formally given names, they were already extinct in Sri Lanka. This indicates that investigating other museums and private collections may reveal more interesting as well as possibly extinct undescribed species. However, both in India and Sri Lanka, some species that were considered extinct have been rediscovered.

Recent studies have identified the six main threats to the amphibians of Sri Lanka. They are habitat loss and fragmentation, pollution, invasive species, pathogens and diseases, road kills and climate change.

Habitat Loss & Fragmentation

This is considered by far the most major threat to Sri Lanka's amphibians. The major contributors to this threat are the human-mediated clearing of natural habitats for the expansion of agricultural lands, plantations (tea, coffee and other crops) and human settlements, hydropower projects and timber harvesting. Sri Lanka has lost more than 70 per cent of its natural forest cover during the last 150 years, and the current rates of depletion of forests and wildlife habitats are considered to be one of the highest in South Asia. The majority of the endemic species of amphibian are restricted to lowland and montane rainforests, where the habitat loss has been most severe. Increased human settlements around important ecosystems (for example in Sinharaja, Knuckles, Kottawa, Kanneliya, Peak Wilderness and Nilgala) would further exacerbate the pressure on these natural habitats. Further, as a consequence of rapid urbanization and industrialization, many marshlands and paddy fields have been reclaimed and drained, thus reducing or completely obliterating ideal habitats for aquatic amphibians.

Pollution

The prevalent levels of the application of agrochemicals, especially in rice fields, and vegetable and tea plantations, have increased over the past three decades. Similarly, the release of untreated industrial wastewater to natural water bodies has intensified. As a consequence, many streams and canals have become highly polluted. The use of pesticides directly decreases the insect population, an important source of food for amphibians. Furthermore, these pollutants can easily make the water in paddy fields and the insects on which the amphibians feed toxic or increase the nitrogen content of the water. The highly permeable skins of amphibians would certainly cause them to be directly affected by these

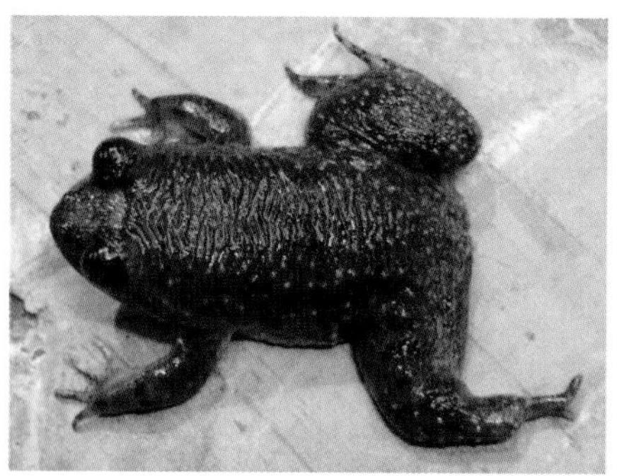

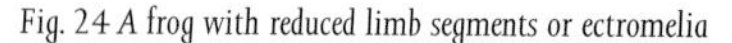

Fig. 24 A frog with reduced limb segments or ectromelia

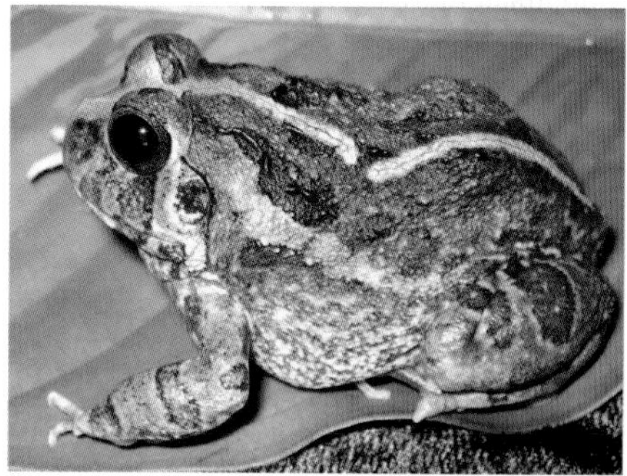

Fig. 25 A frog with lateral deviation of the spine or scoliosis

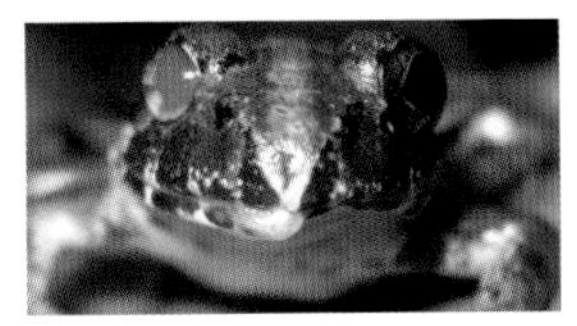

Fig. 26 *A frog with a distorted jaw* (Brachygnathia) *and an eye infection*

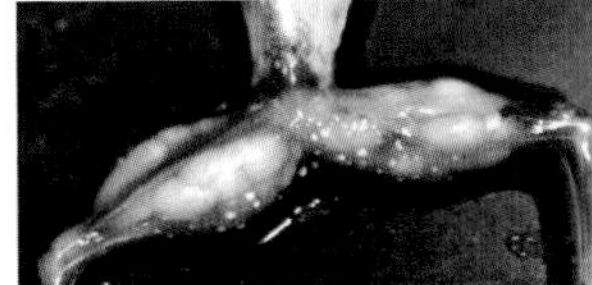

Fig. 27 *A frog with a parasitic* (*cestode*) *infection*

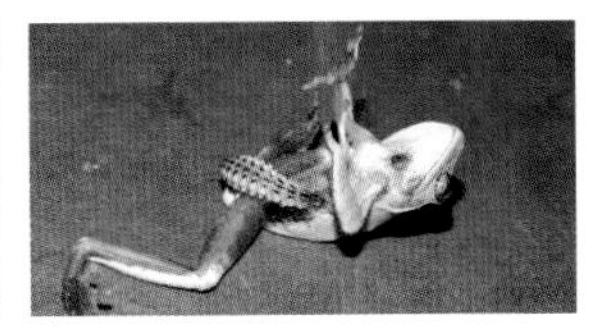

Fig. 28 *A frog being attacked by a beetle larva*

chemicals. This could be one of the reasons for the decline in the populations of some frogs (such as *Minervarya agricola*, *M. greenii* and *Lankanectes corrugatus*) that were commonly found in the paddy fields of the country two decades ago. Similar observations have been made in India. Furthermore, recent studies indicate that tadpoles exposed to pesticides are highly likely to be infected by parasites, leading to malformations. The recent detection of malformations, abnormalities (Figs 24–26) and parasitic infections (Figs 27 & 28) in frogs and toads of Sri Lanka might be indicating the effects of pollution on amphibians.

Invasive Species

The introduction of alien invasive species (AIS) has been identified as one of the major reasons for the decline of amphibians globally. This has not yet been a significant threat to Sri Lankan amphibians. However, the guppy fish, *Poecilia reticulata*, which was introduced to Sri Lanka by the anti-malaria campaign as a mosquito larvivore between 1928 and 1945, has been observed to feed on the tadpoles of *Uperodon* species and *Polypedates cruciger*, but not of *Duttaphrynus melanostictus*. The guppy, which is originally from the neotropics, is now well established in Sri Lanka and can be seen in large populations in paddy fields and associated streams in many parts of the country. Therefore, its effects on the native amphibians should be further investigated.

Pathogens and Diseases

Among diseases, chytridiomycosis caused by the fungus *Batrachochytridium dendrobatidis* has been responsible for rapid population declines and extinctions in many amphibian species globally. Although the disease has been detected among frogs in certain montane locations in Sri Lanka, it is not known whether it has caused any rapid declines in amphibian species. However, it is important to continuously monitor for the presence and prevalence of the disease in Sri Lankan frogs.

Road Traffic Fatalities

Amphibian mortality due to road traffic is a widespread problem globally that has been known to be responsible for population reductions and even local extinction in certain instances. In Sri Lanka, amphibian mortalities due to road traffic are highly prevalent on roads that go through paddy fields, wetlands and forests. Further, they are especially intensified on rainy days when amphibian activity is high. Recent studies indicate that amphibian road kills are exacerbated in certain national parks in the country due to increased visitation. According to recent estimates, several thousand amphibians are killed annually due to road traffic (Fig. 29).

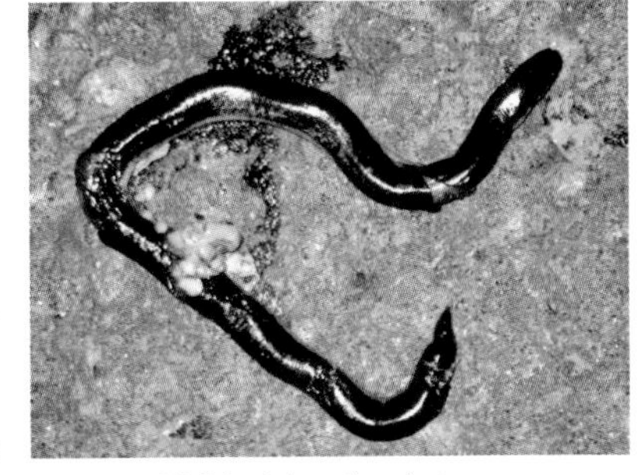

Fig. 29 *A road-killed* Ichthyophis glutinosus

Climate Change

Amphibians are animals that are highly sensitive to changes in environmental temperature, humidity and rainfall, as these factors are critically important for their survival and the completion of their life cycles. Long-term changes in these factors are documented globally and are termed climate change. This is very likely linked to anthropogenic activities and is known to cause changes in rainfall and intense droughts leading to desiccation of aquatic habitats, which are essential for amphibians. A trend in increasing temperatures and decreasing annual rainfall has also been documented in Sri Lanka. It is believed that such changes are due to the clearing of large tracts of tropical forests in the recent past in the Central Highlands. Forest dieback that has been observed in the montane forests of Sri Lanka is also thought to be linked to climate change to a certain extent. Although nothing is known about how such changes have affected Sri Lankan amphibians, it is important to investigate how future changes in the climate are going to affect them.

Amphibian Numbers Affected

According to the IUCN Red List of Threatened Species, 72 (60 per cent) of amphibian species in Sri Lanka are threatened with extinction due to the aforementioned reasons, either directly or indirectly. Eighteen species are Extinct and 17 are Critically Endangered (Table 3). There is therefore an urgent need for continued population monitoring, combined with field surveys, to identify the factors that threaten Sri Lanka's amphibian populations.

Conservation Measures

Despite the high diversity of amphibians in Sri Lanka, it is alarming to note that the majority of the amphibians in Sri Lanka are threatened with extinction (Table 3). However, the discovery of 14 new amphibian species during the past 10 years (2010 up to December 2020) highlights the fact that the diversity of Sri Lankan amphibians may not yet be fully estimated. Furthermore, many of these new species are known only from a few individuals or small populations living in fragmented and isolated habitats, and by virtue of their small size they are very vulnerable to the threatening processes described. However, four of the species that were thought to be extinct (*Adenomus kandianus*, *Pseudophilautus semiruber*, *P. stellatus* and *P. hypomelas*) were recently rediscovered, bringing the number of extinct species to 18.

Sadly, at the time of writing minimal efforts have been made by government management agencies to conserve the Critically Endangered and Endangered amphibian

Table 3. Threat Status of Sri Lanka's Amphibians

Threat status	No. of species	Percentage (%)
Extinct	18	15
Critically Endangered	17	14.2
Endangered	38	31.7
Vulnerable	17	14.2
Near Threatened	04	3.3
Least Concern	20	16.7
Not Evaluated	06	5.0

Source: IUCN Amphibian Red List evaluation report, 2020

species. For most amphibians of Sri Lanka, limited studies have been conducted on their natural history and captive breeding. Thus, very little information is available on these important aspects, which is necessary for the implementation of conservation strategies. Moreover, even the little knowledge available is not being applied for conservation. Furthermore, it was not until 1993 that legal protection for amphibians was initiated by an amendment to the Fauna and Flora Protection Ordinance (Section 31A, Schedule 111), according to which all amphibians of Sri Lanka are now legally protected, making it prohibited to capture or kill any species.

Habitat loss and fragmentation of remaining habitats are highly detrimental to all threatened species of amphibian, especially the ones with very restricted ranges. Although most of the important habitats are protected and managed by the Department of Forest Conservation and Department of Wildlife Conservation, illegal encroachment and clearing still pose a significant threat to these pristine habitats. Given the critical importance of maintaining these highly important areas, it is the duty of all citizens to report any anthropogenic changes that are observed in them to the relevant authorities.

Although the majority of Sri Lanka's amphibians require pristine natural habitats, some are able to exploit man-made habitats, such as home gardens, paddy fields, irrigation canals and man-made reservoirs, or 'tanks'. Both endemic species (such as *Lankanectes corrugatus*, *Minervarya greenii*, *M. kirtisinghei*, *Hydrophylax gracilis*, *Microhyla mihintalei*, *Uperodon rohani*, *U. obscurus*, *Pseudophilautus popularis*, *P. rus* and *P. regius*), and non-endemic species (like *Duttaphrynus melanostictus*, *D. scaber*, *Microhyla ornata*, *Euphlyctis cyanophlyctis*, *E. hexadactylus* and *Hoplobatrachus crassus*), are known to occur very commonly in home gardens, paddy fields and plantations. Dairy farms in the wet zone are also known to attract and harbour caecilians, as they are drawn to the earthworms and insects that live in decomposing cow dung. Moreover, if a well-shaded, ecofriendly home garden is maintained, amphibians will continue to thrive in this anthropogenic habitat. Ornamental ponds also help them with breeding, as some species require water to lay eggs and for their tadpoles to develop. Reducing the use of inorganic fertilizers, pesticides and insecticides that are toxic to amphibians will ensure their continued survival in these habitats.

Understanding the critical ecosystem services that amphibians provide is also of crucial importance. Amphibians play an important role in controlling insect pests, as they are predominantly insectivorous. Educating the public about this importance would be helpful in justifying the importance of conserving amphibians. Furthermore, eradicating misconceptions about frogs – such as that tree-frogs are poisonous to humans – through awareness is also important, because tree-frogs are killed regardless in certain parts of the country. This would be highly beneficial for people living on the buffer zones of wildlife habitats who frequently come across amphibians. Thus, awareness programmes to spread the word about the importance and conservation of amphibians are a need of the hour.

In Sri Lanka during the past 15 years, several small steps have been taken towards the development of necessary knowledge and tools for the conservation of amphibians. A few ex-situ conservation breeding programmes and releases of some amphibian species have occurred. Some species, such as *Uperodon obscurus* and *Polypedates cruciger*, which had meagre populations, have now increased in some areas where ex-situ breeding programmes were carried out. In addition, some commendable work on the captive breeding of direct-developing amphibians has been carried out. This information would be highly beneficial should captive management be required as a last resort of species conservation. The Amphibian Specialist Group of Sri Lanka of the IUCN Species Survival Commission has recommended the following conservation measures:

- Investigating the pattern and incidence of malformations, abnormalities, injuries and parasitic infections in amphibians of Sri Lanka.
- Initiating captive breeding programmes for endangered species where in situ management is not possible.
- Carrying out a rigorous analysis of amphibian extinctions in Sri Lanka.
- Initiating immediate steps to reduce amphibian road kills.
- Formulating a National Management Plan and a monitoring programme for amphibians of Sri Lanka that would be implemented by the Ministry of Environment and the Departments of Wildlife Conservation and Forest Conservation.
- Conducting awareness programmes among the residents of the buffer zones in important wildlife habitats.
- Encouraging farmers to employ organic farming methods through the provision of incentives.
- Studying traditional beliefs and practices regarding amphibians to enable the utilization of some of these beliefs in public awareness programmes and conservation.
- Introducing a module on reptile and amphibian diseases and treatment into the undergraduate veterinary medicine curriculum.

As with the conservation of biodiversity in many other countries, a major problem inhibiting this achievement in Sri Lanka is the lack of sustained investment in research and the applied management practices that must follow. Conservation is often seen as a luxury activity and one that falls far down in the ranks of national spending or international aid provisioning. However, society in general is beginning to understand the economic value of ecosystem processes such as clean drinking water, soil stability and the effects on local climate, and the appropriate management of wilderness areas and human-altered systems where natural processes and biodiversity persist.

Amphibians in Archaeology, History & Traditional Medicine

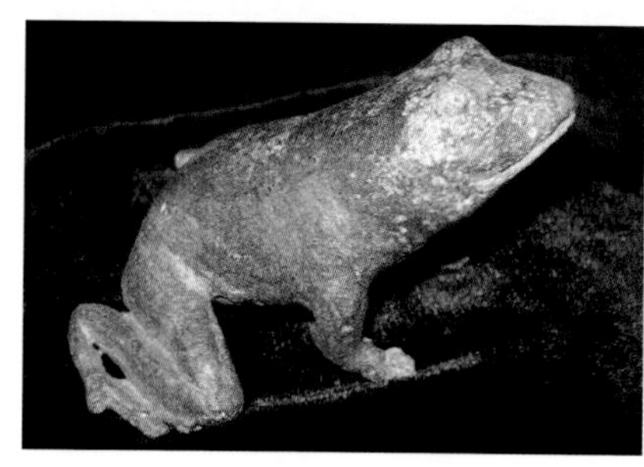

Fig. 30 Sixth-century AD bronze cast of a frog

There is much archaeological, historical and literary evidence to show that amphibians have attracted the attention of the people of Sri Lanka from ancient times. This is especially evident from the discovery of an ancient bronze cast of a frog (Fig. 30) from excavations conducted by the Department of Archaeology and the Central Cultural Fund in Anuradhapura.

Stratigraphic evidence from excavation sites indicates that these objects belong to an era between the sixth and eighth centuries AD (Anuradhapura and Jetavanãrãma Museum records). The bronze cast had been inside a special 'relic box' along with casts of a crab, fish, water snake and freshwater terrapin, and an effigy of the goddess Lakshmi. The relic box was placed inside a '*weva*', one of the hundreds of irrigation tanks built by ancient kings, provincial rulers, ministers and commanders of the royal army. It is possible that this ritual was connected to a belief that these five aquatic animals kept the aquatic ecosystems clean. Furthermore, a unique and accurately made solid-gold cast of a frog (probably *Euphlyctis cyanophlyctis*, the skipper frog), which weighs 41.9g and has a snout to vent length of 38mm (Fig. 31) was also discovered in Anuradhapura. It is possible that a systematic search of all tanks will unearth more artefacts of archaeological value and herpetological interest (although it is also possible that treasure hunters may already have robbed most of these artefacts).

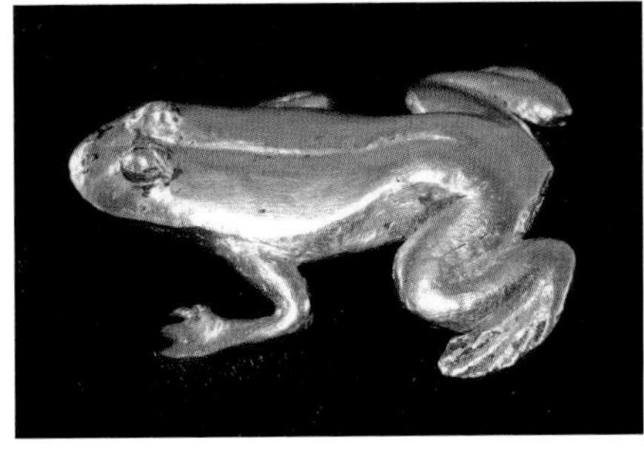

Fig. 31 Sixth-century AD gold cast of skipper frog

Ancient Coins Depicing Frogs

Indian coins (Pallava, Chola) which were minted around 500 BC, depicting various figures (such as turtles, fish, lions, elephants, tigers, horses, cattle, deities, kings, foliage and frogs), have been in circulation in Sri Lanka (Seyone, 1998). Coins that depict frogs and include other figures have been discovered in many places in the country (Fig. 32).

Early Literary Records of Amphibians

Mahãwamsa, the great chronicle of Sri Lanka, records that King Dhammãsoka of India worshipped the great bodhi tree (*Ficus religiosa*) daily, and his jealous wife, Queen Tissarakkhã, caused the tree to perish by using toad poison. However, some authorities do not agree with Geiger on this interpretation.

One of the first medical works to originate in Sri Lanka, the *Sãrartha Samgrahaya* (340–368 AD), suggests that the traditional management of snakebite envenoming and toad poisoning had been in existence in Sri Lanka for nearly 2,000 years. The *Yõgãrnavaya*, written during the reign of King Buvanekabã I (1272–1284) records the administration of *Devun diya* (a medicinal drink) for poisoning due to an amphibian bite.

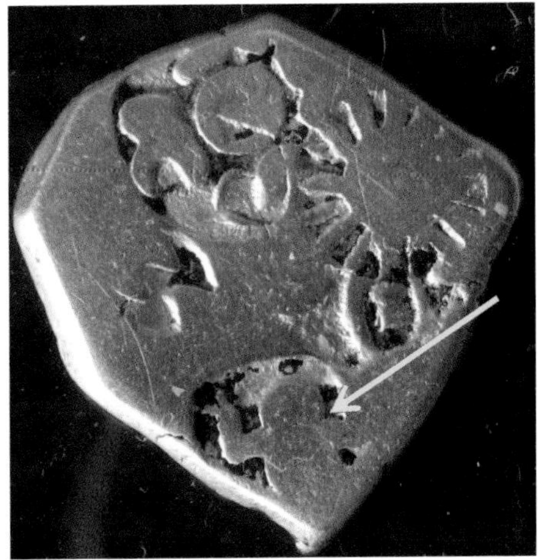

Fig. 32 Silver coin of Indian origin depicting a frog

Ancient Place Names Referring to Frogs

There are more than 150 village and pond names associated with frogs dating from the time of the early Sinhalese kings. For example, *Bekkegama* ('*bekke*' = paddy field frogs, '*gama*'= village), *Maduvila* ('*madu*' = frog, '*vila*'= pond). Other examples include *Deduru Oya* (frog stream) and *Mediyawa* (frog lake). It is reasonable to assume that the early inhabitants would have understood the ecological role played by amphibians in agriculture, their association with water and the benefits of them preying on insect pests.

History of Amphibian Studies

According to the above, we can see that Sri Lankans were well aware of the amphibians living around them, but the scientific study of the amphibians of Sri Lanka began with the arrival of Europeans, especially when British colonists sent specimens abroad to the Natural History Museum, London.

The first amphibian from Sri Lanka to be featured in the scientific literature was *Ichthyophis glutinosus*. This was published as *Serpens Caecilia Ceylonica* by Albertus Seba in his *Thesaurus* in 1735. Carolus Linnaeus gave the specific name *glutinosus* to it in 1758. Subsequently, many species of amphibian were described by foreign scientists working in museums, including George Albert Boulenger (1858–1937), Albert Günther (1830–1914) and Johann Gottlob Schneider (1750–1822).

Edward Frederick Kelaart, who described *Polypedates schmarda* (now *Pseudophilautus schmarda*) in 1854, can be considered to be the first Sri Lankan to describe an amphibian from the country. Before this Kelaart featured nearly 10 known species of Sri Lankan amphibians in his book, *Prodromous Faunae Zeylanicae*. However, *The Amphibia of Ceylon*, written by Parakrama Kirtisinghe in 1955, was the first book to be dedicated to the amphibians of Sri Lanka, and it also had the distinction of being the first book to be written by a Sri Lankan on the subject. Subsequently, many local herpetologists and amateurs were involved in describing and studying the diversity and life history of Sri Lanka's amphibians, which vastly increased our knowledge of this unique group of animals.

Amphibians in Traditional Medicine

Described below are some of the important medicines (mainly medicinal plants) used in the traditional treatment of poisoning due to amphibians, and also the preparation of these medicines as documented in the literature. These may benefit future workers on medical anthropology, pharmacology and traditional treatment. According to traditional systems, the collection of medicinal plants is ritualized. They may be picked only at particular times of the day and only in certain months. Rules govern the method of even breaking the needed part, the places recommended for collection and so forth.

Most of the traditional books on this subject are out of print or difficult to obtain. Most are written as palm-leaf (*ola*) manuscripts, which are not commonly available and difficult to access by the majority of people. However, it should be noted that to date there have been no authentic records of severe cases of poisoning in humans due to amphibians in Sri Lanka. Furthermore, little is known about the toxins of Asian amphibians, nor have any of the traditional treatments been scientifically proven to neutralize amphibian toxins. As some amphibian species are often found inside houses, there is frequent exposure to amphibian toxins when the animals are trampled upon or accidentally touched, or due the sudden squirting of amphibian urine on to the human body.

The Ayurvedic medical system is composed of eight branches known as *astānga āyurveda*, of which *Agadatantra* (*Agada* = anti-poison remedy and *tantra* = science) is the one that deals with the treatment of poisoning from plants, animals and other elements or compounds. *Jangama visha* (animal toxins) is one component of *Agadatantra*, which includes the management of poisoning by amphibians.

According to traditional literature, tautness, tenderness, swelling and yellowing around the site of a bite/contact are signs and symptoms of amphibian poisoning. At the same time, the victim is supposed to feel thirsty, nauseous and drowsy, and to suffer from diarrhoea, with the body becoming cold. It is stated that if the person is not treated properly they may fall ill again.

The under-mentioned concoctions and techniques were taken from published literature, palm-leaf manuscripts, and *Ath poth* or handwritten reference manuals. The following is a translation (by the senior author) of verses from the *Ath potha* of the late E. D. Jamis Appuhamy, a well-known traditional physician of Sri Lanka. The word 'bite' used in these verses may mean contact with skin toxins of amphibians.

Verse 1 When a poisonous frog bites, the bite site becomes taut, yellow and tender with swelling. Other features include thirst, vomiting and drowsiness.

Verse 2 It is necessary to treat the bite of an *ätikitta* (*Pseudophilautus* species) immediately. If the victim is not treated properly, body ailments could develop later.

Verse 3 The body becomes cold when a frog bites. It disturbs the bile and diarrhoea sets in. Provide the victim with the ground bark of the *rath hańdun* (*Pteroarpus santalinus*) tree with lime juice to drink to stop diarrhoea.

Verse 4 If a tree-frog jumps and scratches the body, do not repent later by not resorting to medical assistance. The victim will become thin; hence apply gingili oil on the body first. Repeat this for three days and bathe the victim on the fourth day. Give *kūra tampalā* (*Amaranthus viridis*) porridge to drink for neutralizing the poison.

Verse 5 Leaves of *pusväl* (*Entada phaseolodes*) are pounded with raw turmeric and salt and cooked over steam. Foment the body with this cooked mixture and tie the residue over the bite to neutralize the local reactions and the poison.

The following prescriptions are from the palm-leaf manuscripts in the National Museum (de Silva and Wickramaratne, 1956): from *Sarpa Veda Pota* (snakebite medicine) (Colombo Museum Library no. A K 10):

1. Boil down to a paste the kernel of *kapu* (*Gosypium herbaceum*) seeds, the kernel of *labu* (*Citrullus colocynthis*) seeds, and *kalanduru* (*Cyperus rotandus*) in coconut oil and apply to the bite wound for *geyi gembā* (*Duttaphrynus melanostictus*) poisoning.
2. Boil *atamburu* (*Fagraea ceilanica*) leaves and apply on the body.

The following are a few examples of prescriptions for frog and toad poisoning from the first book in the English language on the traditional treatment of snakebite by Emmanuel Roberts.

1. An aqueous extract of fresh *kūra tampalā* (*Amaranthus viridis*) roots, stems and leaves is given internally in doses of 1–2 ounces.
2. The fresh latex (milk) of *varā* (*Calotropis gigantea*) is applied to the wounds.
3. The crushed fresh *murungā* (*Moringa pterygosperma*) leaves are applied to the wounds inflicted by frogs.

Prescriptions and remedies for amphibian poisoning are included in many traditional medical books on the treatment for snakebite. The following are a few examples.

1. Apply the juice of *ratu pitawakkā* (*Phyllanthus urinaria*) leaves mixed with coconut milk on afflicted sites of the body. Extract juice from a whole *kuppamēniā* (*Acalypha incdica*) plant. Cook this in tana *hāl* (*Setaria italica*) and give the broth to drink.
2. Gunasena (1963) records the following: if the body of a patient suffering from frog bite becomes cold and diarrhoea sets in, administer ground rat *hańdun* (*Pteroarpus santalinus*) with lime juice orally. Also, apply clarified butter mixed with coconut milk on the wound.

Amphibians in Folklore

There is evidence that some geoponical (slash-and-burn agriculture) societies existed in Sri Lanka from c. 17,500 calibrated years BC. This is considered the earliest form of agriculture so far discovered in Sri Lanka, as well as in southern Asia. Agriculture was irrigated by small lakes to large man-made tanks, even around the third century BC. This 'tank civilization' existed in close association with water bodies, agricultural fields, homesteads, forests and temples. Agricultural workers would thus have inevitably encountered a wide variety of amphibians. They would have heard their varied calls, seen them feeding on insects and observed the commensal habits of some toads (such as *Duttaphrynus melanostictus*) and frogs (like *Minervarya* sp. *Polypedates maculatus*). These early inhabitants would also have experienced instances of mild poisoning by some amphibian toxins. The existence of many beliefs, practices and folk tales woven around amphibians in Sri Lanka is thus not surprising.

It is of great interest to note that some beliefs and traditions regarding amphibians given below indicate that humans have attributed supernatural powers to them. From the sixth century AD, frogs have been figured in art as a symbol of water purification. They were also considered as animals of good fortune. Furthermore, antidotes for poisoning by amphibians have been documented in traditional medical works from the fourth century AD (see page 31). The Buddhist story of the *Mandūka devaputta* records that the Bodhisattva (a previous incarnation of Prince Siddhartha) was born as a frog in *Haritamāta Jātaka*.

The impact of amphibians on the ancient society is clearly seen from the fact that many irrigation tanks (*väv*), ponds (*pokunu*) and villages (*gam*) were named after amphibians (see page 29).

Various Sri Lankan traditions, proverbs and folklore about amphibians are summarised here. These were collated from Buddhist priests, village elders, traditional snakebite physicians and astrologers from various parts of the island and from the available literature. This list is by no means comprehensive. The beliefs are listed under the following categories: those that feature the 'good' qualities of frogs and their association with nature, and those that feature the poisonous, evil and mysterious aspects of frogs.

'Good' Qualities of Frogs & Their Association with Nature

These beliefs have positive effects on the conservation of amphibians.

1. Some believe that seeing a frog or toad when leaving on a journey presages a good journey.
2. It is believed that dreaming of frogs indicates that the dreamer will be blessed with children or riches.
3. Toads that reside in houses are believed to be protectors of buried treasure.
4. Due to the commensal habits of *Duttaphrynus melanostictus*, which hides in cool, dark places inside human dwellings during the day, many villagers call it *Gei niväsi* (*Gei* = house and *niväsi* = resident). Some call it *Gei poloňga* (*Gei* = house and *poloňga* = viper). Some tolerate its presence and allow it to live in its usual niche inside the house.
5. Terrestrial activities of tree-frogs are believed to indicate a drought.
6. Some believe that the absence of frogs and toads in agricultural fields is an indication of impending crop failure.
7. A parallel for the above belief is that there are no frogs on infertile land or water bodies.

Poisonous, Evil & Mysterious Aspects of Frogs

These beliefs have negative effects on the conservation of amphibians.

1. Toads have a bad reputation for injecting poison into a wound inflicted by their bite.
2. Parents instruct their children not to kick toads as they could be envenomed by the

cloacal end of the toad (de Silva, 2001).

3. Some believe that if a tree-frog (genus *Polypedates*) leapt on children, they would become consumptive and attenuated like that frog. Furthermore, it is assumed that if a tree-frog jumps on to a body and scratches, the victim should be treated immediately to neutralize the poison.
4. Some believe that merely touching a tree-frog may result in being infected with leucoderma (development of pale patches on the skin).
5. When some tree-frogs jump, they squirt urine as a defensive mechanism, and many believe that if this urine falls on the body, a skin infection will set in.
6. It is believed that if a toad is imprisoned even under a heavy container, it can get out by 'some mysterious power' without upsetting the container (Simion, 1954).
7. If a tree-frog is seen inside a house at a time when there is a domestic problem, that misfortune is attributed to that particular frog by some people.
8. A parallel for the above belief is that when a tree-frog comes into a house it presages a period of misfortune.
9. Those who dabble in witchcraft consider the house toad *Duttaphrynus melanostictus* to be the vehicle used by departed inmates of the house to get about.
10. It is believed that if a person dies with a craving for their wealth, they will be reincarnated as a toad in the next birth.
11. Some believe that if the house toad *Duttaphrynus melanostictus* is seen in your compound, that particular place has buried treasure, and that a demonic spirit in the form of a toad has come to protect the treasure. If these toads are harmed it will have serious repercussions.

Proverbs

The following proverbs, though common in day-to-day life, are taken from Seneviratne (1936):

1. Even in the lotus pond there are frogs (that is, you find the vulgar even in a good society).
2. However high the water may rise, it will only be up to the neck of a frog.
3. Like collecting frogs into a *lāha* (a *lāha* is a wide-mouthed basket used to measure rice, and when frogs are collected into this receptacle they will inevitably jump out). This signifies an impossible task.
4. Like a frog in a well (that is, sees and knows nothing of the world outside).
5. The following was recorded by Simion and Wijetilleke (1965): 'like the gold frog that was melted' (that is, how a goldsmith cheats on an old woman by saying that her gold trinket frog happened to jump and escape, signifying the deceitfulness of the goldsmith).

Amphibians in Folk Tales

Kavata Andarē, the well-known royal court jester of Sri Lankan folklore, lived during the reign of King Rajadhi Rajasingha (1780–1798 AD) of Kandy. There are many interesting and hilarious fables spun around Andarē, most of which are popular among Sri Lankans.

According to one fable, the incessant croaking of a frog disturbed the sleep of King Rajadhi Rajasingha. After many sleepless nights, the King was reduced to a state of great distress and felt that he could no longer put up with it. Andarē was the obvious choice to solve the problem. The royal court was unanimous in supporting the King in this choice – no one else but Andarē was the man for the job. Court messengers were swiftly dispatched to inform Andarē of the King's orders, and he hurried to the Royal Palace armed with his bow and a quiver-full of arrows, while racking his brains for a plan of action. He questioned the King on the whereabouts of the offending frog, but alas, the King could only point vaguely in the direction from which the sound appeared to come. Andarē selected an arrow, set it on the bow-string, drew the bow to its fullest and sent the arrow on its way in the direction the King indicated.

By some miracle, the croaking of the frog was not heard thereafter. Andarē had done it again! The King was jubilant; and so was the Queen and the rest of the royal household. All marvelled at Andarē's skill. As a reward, the King bestowed upon Andarē and his descendants the *pelapat nāmaya* (family name) starting with *Sadda Vidda* (shot the sound). To this day, Andarē's descendants use this surname – Sadda Vidda Palangu Pathirana Hewathandilige Andrē Muthu Kumaran.

Searching for Amphibians in Sri Lanka

Given the tropical humid conditions of the island, amphibians are abundant in Sri Lanka and can be found in a plethora of habitats. If you live in a rural or urban area, you can start searching for them around your house – common species, such as *Duttaphrynus melanostictus*, *D. scaber*, *Uperodon taprobanicus*, *U. rohani*, *U. obscurus*, *Polypedates maculatus* and *P. cruciger* tend to occur in and around human dwellings (in washrooms/bathrooms) and compounds. Occasionally a few *Pseudophilautus* species (usually *P. popularis* or *P. regius*) may be encountered in home gardens. They usually hide during the day under flowerpots, leaf litter, rubble and decaying logs in gardens. Ornamental ponds in gardens are also ideal places to look for amphibians if you have one in your garden. Another ideal place to look for them is paddy fields. These are excellent for finding aquatic and semiaquatic species including *Minervarya agricola*, *Euphlyctis cyanophlyctis* and *Hydrophylax gracilis*.

It is also easy to see frogs and toads on roads immediately after heavy rains even in the dry zone. The best thing to do is to drive slowly along roads immediately after rains, and after dusk. Similarly, frogs congregate and vocalize in roadside puddles, also after rains. However, the best time to see amphibians is during the monsoon seasons. Other ideal habitats for seeing frogs are along forest paths and streams, and around tanks (man-made lakes) and ponds after dusk.

Although amphibians are common in forests and rainforests in the wet zone, the best time to see them is also after 6 p.m. during the rainy season. Rather than carry a torch, it is better to use a powerful, well-focused headlamp that leaves both your hands free for photography (but note that you need permits from the Department of Wildlife Conservation to capture amphibians). Always wear protective footwear to avoid the possibility of snakebite. When working in some areas of the dry zone scrub forests, be cautious of wild elephants. Carry a snake stick or similar object to use to disturb the forest leaf litter, and turn over logs or stones. Always turn logs and stones towards you, so that if there is a snake hiding underneath a log, it cannot strike at you, as the log or stone will act as a protective barrier. While walking along forest paths, gently tap low bushes with a stick, as many shrub frogs that perch and hide on low vegetation will then move, revealing their presence. Decaying logs may also be opened since amphibians usually hide inside such places. You could also remove dead tree bark and check tree-holes and crevices, as these are good hiding places for many amphibians. For larval amphibians it is best to check small pools, slow-moving streams and temporary puddles. Tadpoles could be collected by using a dip net, then observed in tanks. Always remember to adhere to strict biosecurity measures (for example, disinfect boots and the like before entering a critical habitat or forest). For further details on field techniques, refer to de Silva & Mahaulpatha (2007). Make sure you release any tadpoles collected back into their natural environment.

Glossary of Technical Terms

adult Sexually mature stage.
amplexus Act of mounting and clasping of female tightly by male frog with forelimbs during copulation.
annuli Grooves on skin that run along body in caecilians.
aquatic Living in water.
arboreal Living above the ground, for example on trees.
auxiliary amplexus Type of amplexus that involves gripping female just behind forelimbs.
batrachology Study of amphibians.
calcar Pointed appendage on heel or elbow of frogs.
canthal ridge The ridge that extends from anterior mid-border of eye to naris, which separates lateral from dorsal aspect of snout.
clutch Total number of eggs laid by female at one time.
dorsolateral glandular fold Distinct swollen ridge of glands, which starts behind eye and ends at groin.
endemic Naturally restricted to particular geographic region.
endotrophic Obtaining nourishment from yolk of egg, usually seen in direct-developing species of amphibian where larvae obtain nourishment for development from yolk.
fossorial Refers to species that live for the most part inside humus or soil.
herpetology Study of amphibians and reptiles.
imago Newly metamorphosed adult form.
inguinal amplexus Type of amplexus where male holds female near hindlimbs.
interorbital space Region on head between orbits (eyeballs).
lingual papilla Bony, tooth-like structure in lower jaw.
longitudinal ridges Distinct raised longitudinal ridges on dorsal and lateral aspects of body.
loreal region Area between eyes and nostrils.
monotypic Refers to a particular genus that consists of only one species.
parotid glands Two distinct glands situated on head behind eyes; contain toxins.
serrated Possessing saw-tooth edge.
shagreened Refers to skin covered with minute granules.
shank Leg below knee down to foot.
spawn Eggs of aquatic animals (here, the term is specifically used for eggs of amphibians).
spinules Conical-shaped, papilla-like structures on body's surface.
supratymphanic fold Distinct ridge that starts from behind eye and runs above tympanum.
terrestrial Living primarily on land.
tubercle Knob-like projection.
tympanum External ear-drum.
webbing Skinny membrane between toes that is usually well defined in aquatic frogs.

Species Descriptions

The following section contains descriptive accounts of all 102 extant species of amphibian in Sri Lanka. They include the current names (including the scientific and common names), historical aspects, identification features, habits, habitat, distribution, taxonomic status, colour photographs and IUCN Red List categories. Brief descriptions, along with the common names and historical aspects, are given for the 18 extinct species. Species descriptions are organized in amphibian orders, families and genera. Species common names are given in English, and in Sinhalese in parentheses below. Sizes for frogs and toads are abbreviated to SVL (snout to vent length), and for caecilians to TL (total length).

ORDER Anura (Frogs & Toads)
Anurans are characterized by the presence of four limbs, a stout body, protruding eyes, limbs that are kept folded underneath the body, tongue that is attached anteriorly in the mouth and absence of a tail. They typically lay eggs in water, where the eggs hatch into larvae known as tadpoles, which gradually metamorphose into adults. Some species complete the aquatic larval stage within the egg and hatch into miniature adults, a phenomenon known as direct development. Adult anurans are carnivorous, while most of the tadpoles are herbivorous. About 7,300 species in 54 families are known from all over the world except the Antarctic. The Sri Lankan amphibia is represented by 117 species of anuran in the six families Bufonidae, Dicroglossidae, Microhylidae, Nictybatrachidae, Ranidae and Rhacophoridae.

Family Bufonidae (Toads)
Members of the family Bufonidae are characterized by thick, warty, dry or glandular skin, and shortened forelimbs and hindlimbs used for walking or hopping. There is a concentration of glands in the temporal-neck area forming the prominent parotid gland, which stores an alkaloid secretion. The bufonids are generally called toads. Adult toads lack teeth in the upper jaw. The pupil of the eye is horizontal. They reproduce by depositing strands of eggs in water and the eggs hatch into tadpoles. The tadpoles have keratinized mouthparts. In most species fertilization is external, but a few are known to fertilize internally. The family Bufonidae is distributed worldwide except Antarctica, Australia and Oceania. More than 630 species are known throughout the world, which belong to 52 genera. Sri Lanka is home to six species of bufonid comprising two genera, *Adenomus* and *Duttaphrynus*. The genus *Adenomus* is endemic to Sri Lanka.

Genus *Adenomus*
Adenomus species are characterized by their slender build and moist, warty skin. They have comparatively long limbs with smooth finger edges. However, they lack the supraorbital ridge, which is prominent in *Duttaphrynus* species. Two species of *Adenomus* are found in Sri Lanka.

Kandyan Torrent Toad *Adenomus kandianus*

(Nuwara kuru gembã)

First described as *Bufo kandianus* by Albert Günther in 1872. The species was not reported after 1872 and was considered extinct in 2004. However, in 2012 it was rediscovered in the Peak Wilderness Sanctuary. *Adenomus dasi* is considered to be a junior synonym of *A. kandianus*.

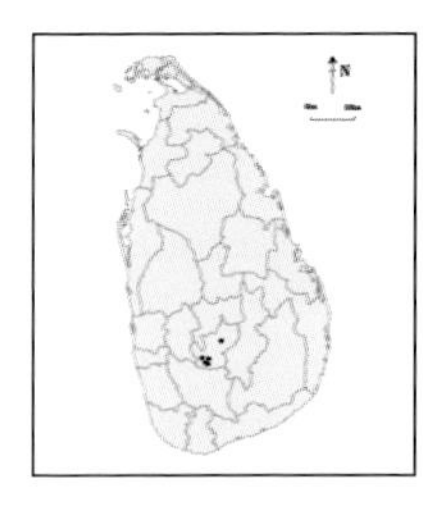

Size SVL 30–45mm.

Identification Features Body slender, flat with elongated limbs and smooth finger edges (Fig. 1). Dorsum moist with small, rounded and spinous tubercles. SVL of tympanum not visible. Scattered small, spiny warts on dorsal surface of body. Toes partially to fully webbed. Distinguished from *A. kelaartii* by absence of cranial ridges in *A. kandianus* and presence of relatively long and narrow parotoid glands (vs shorter parotoid glands in *A. kelaartii*).

Colour Dorsal side a mixture of light brown, orange and yellow (Figs 1 & 2). Head more reddish relative to body, and parotoid glands chocolate-brown or dark brown. 'X'-shaped marks formed by tubercles are dark orange-brown. Belly a mixture of off white and yellow with brownish patches. Black and gold dots on lips. Gold-coloured vertebral stripe at posterior dorsal side of body (Fig. 2).

Habits Terrestrial, nocturnal species. Closely associated with clear flowing streams, and hiding under rocks, decaying logs and leaf litter along stream banks by day.

Habitat and Distribution Inhabits submontane and montane cloud forests at 1,000–1,879m above sea level. Known from Moray Estate, Rajamally, Maskeliya, Peak Wilderness Sanctuary and Piduruthalagala Forest Reserve. Always associated with clear flowing streams in forests.

Status Endemic.

IUCN Red List Category Endangered.

Fig. 1 Dorsolateral aspect

Fig. 2 Lateral aspect

KELAART'S DWARF TOAD *Adenomus kelaartii*

(Kelartge kuru gembã)

First described as *Bufo kelaartii* by Albert Günther in 1858. Later placed in the genus *Adenomus*.

Size SVL 25–80mm.

Identification Features Body slender, flat with elongated limbs, and smooth finger edges (Figs 1 & 2). Dorsum moist with small, rounded, spinous tubercles. Long, narrow parotid glands on either side of neck behind eyes. Canthal ridge distinct and angular (Fig. 3). No bony ridges on head. Nostrils nearer to tip of snout than to eye. Loreal region concave with bony ridges (cranial ridge). First finger shorter than second one, and tips of digits rounded. Toes webbed and fingers not webbed. Two oval metatarsal tubercles and tarsal ridges present. Outer sides of limbs and venter granular.

Colour Dorsal side a mixture of light to dark brown, and tinge of grey with dark brown pattern symmetrically arranged (Figs 1 & 2). Limbs have cross-bars. In some individuals, lateral aspects have a few red and white patches. Belly white, sometimes with yellow tinge and dark mottling. Eggs, laid in strands, unpigmented.

Habits Terrestrial, nocturnal species, occasionally active by day, and often hiding under boulders, rock crevices and decaying logs, in tree-holes and leaf litter, and along rocky banks of streams. Displays occasional arboreal habits. Toadlets occupy wet leaf litter accumulated on rocks along banks of streams.

Habitat and Distribution Inhabits many discontinuous localities in wet submontane and lowland forests of Sri Lanka (<1500m elevation). Recorded from many lowland wet forest habitats, including Ambagamuwa, Knuckles, Kuruvita, Deniyaya, Haycock, Kanneliya, Hiyare, Kudawa (Sinharaja), Moralla, Palabaddala, Kanneliya and Yagirala.

Status Endemic.

IUCN Red List Category Vulnerable.

Fig. 1 Dorsolateral aspect

Fig. 2 Lateral aspect

Fig. 3 Cranial ridges indicated by arrow

Genus *Duttaphrynus*
Duttaphrynus species toads have a stout and stocky appearance with short legs. Their skin is dry and warty with prominent parotid glands on the head. They are further characterized by the presence of prominent bony ridges, such as canthal, preorbital, supraorbital and postorbital ridges, and short orbito-tympanic ridges on the head region. The snout is short and blunt, the interorbital space is broader than the upper eyelid and the tympanum is small. The first finger extends beyond the second and the toes are half webbed. The finger edges of *Duttaphrynus* are granular (rough). Four species are known from Sri Lanka. These were previously placed in the genus ***Bufo***, but in 2006 they were placed in the new genus *Duttaphrynus* based on molecular evidence. Two species are endemic to Sri Lanka.

Kotagama's Toad *Duttaphrynus kotagamai*

(Kotagamagē gembā)

First described as *Bufo kotagamai* by Prithiviraj Fernando and Nihal Dayawansa in 1994. Later placed in the genus *Duttaphrynus*.

Size SVL 33–63mm.

Identification Features Body short and stout, limbs short and fingers have rough edges. Skin appears dry and warty with numerous tubercles. Head broader than long, and has prominent supraorbital, supratympanic and parietal ridges. Preorbital and postorbital ridges weekly defined. Parotid glands narrow and elongated, and situated dorsally behind eyes. Loreal region concave. Toes webbed. Large tubercles scattered on dorsum, limbs, upper eyelids and edges of both lips (Fig. 3). Outer edges of upper eyelids granular. Can be distinguished from other *Duttaphrynus* species by 'horseshoe'-shaped (inwardly curved) black warty parietal ridge on head (Fig. 1).

Colour Dorsal a mixture of brown, yellow or orange, and grey (Figs 2 & 3). Belly whitish with brown mottling (uneven discoloured patches). Supraorbital and parietal ridges, tips of digits, and tips of spinous warts black. Light cross-band between eyes. Dark cross-band on forearms, forefeet, tarsi and tibias.

Habits Nocturnal species, but in rainforests with a thick canopy, it may be seen active on the forest floor by day.

Habitat and Distribution Mainly confined to lowland rainforest habitats of south-west central Sri Lanka (<1,070m elevation) and known from a few locations such as Sinharaja Forest Reserve, Kitulgala, Samanala Nature Reserve and Massana Forest Reserve.

Status Endemic.

IUCN Red List Category Endangered.

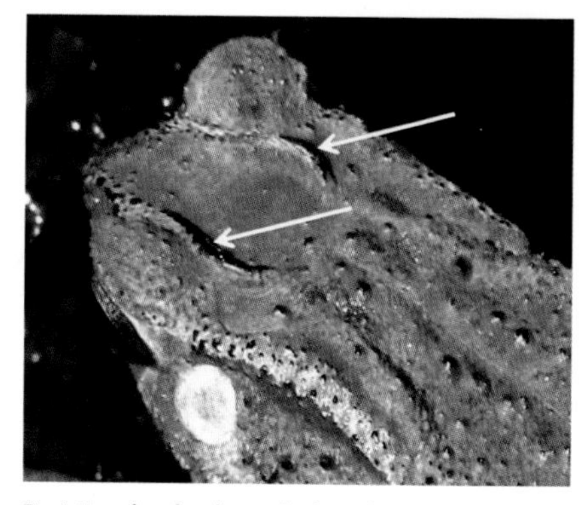

Fig. 1 Horseshoe-shaped parietal ridges indicated by arrows

Fig. 2 Dorsolateral aspect

Fig. 3 Dorsolateral aspect

Common Indian Toad *Duttaphrynus melanostictus*
(Gemba)

First described as *Bufo melanostictus* by J. G. Schneider in 1799. Later placed in the genus *Duttaphrynus*.

Size SVL 50–95mm.

Identification Features Body robust and stout, limbs short and fingers have rough edges. Skin rough and dry with numerous tubercles and warts. Tympanum large and greater than or equal to one-third or more of eye diameter. Interorbital space concave. Nostrils nearer to tip of snout than to eye. Canthal ridge angular. Loreal region concave. Toes webbed and fingers not webbed. Can be distinguished from other *Duttaphrynus* species by large, kidney-shaped, swollen parotid glands, inter-parotoid warts that run up to anterior level of parotid glands and absence of parietal ridges (Figs 1 & 2).

Colour Dorsal side yellowish or brownish, but sometimes brick-red patches also seen on dorsum (Figs 1 & 2). Cranial ridges black. Belly whitish-brown and throat region has brown spots (Fig. 3).

Habits Terrestrial nocturnal species that is occasionally active during the day. Additionally seen in shallow water bodies, including paddy fields at night. Eggs laid in water as long, gelatinous strands that get entangled in the aquatic vegetation.

Habitat and Distribution The most common and widely distributed toad in Sri Lanka, found up to 1,750m above sea level, including in highly disturbed habitats. Commonly found around human

Fig. 1 Lateral aspect

habitations, including houses, urban areas and monoculture plantations, and in degraded forests.

Status Native. Extralimital: India, Pakistan, Bangladesh, Myanmar and Southeast Asia.

IUCN Red List Category Least Concern.

Fig. 2 Dorsolateral aspect

Fig. 3 Ventral aspect

Noellert's Toad *Duttaphrynus noellerti*

(Noellartgẽ gembã)

First described as *Bufo noellerti* by Kelum Manamendra-Arachchi and Rohan Pethiyagoda in 1998. Later placed in the genus *Duttaphrynus*.

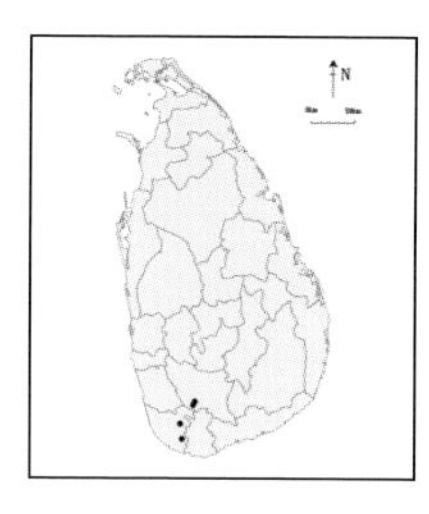

Size SVL 50–89mm.

Identification Features Body robust and stout, limbs short and fingers have rough edges. Skin rough and dry with numerous spiny tubercles and warts (Figs 1 & 2). Can be differentiated from other *Duttaphrynus* species by elongated parotid glands and smaller tympanum and absence of horseshoe-shaped parietal ridge. Body and limbs covered by spinous warts. Head has many ridges including supratympanic ridge that is wider than the others. Sides of head covered with smooth warts with melanic spinules. Resembles *D. melanostictus* and can be differentiated by elongated parotid glands and smaller tympanum (vs swollen parotid gland and larger tympanum in *D. melanostictus*) (Fig. 3).

Colour Dorsum and sides reddish-ash or reddish-brown, marbled with dark brown. Well-defined dark patch runs from behind tympanic region and continues on to flanks and upper surfaces of limbs. Belly off white with brown mottling.

Habits Mainly a nocturnal species. Diurnal activities are known around stream banks with rocks and on the forest floor in rainforests with dense canopy shade.

Habitat and Distribution Mainly confined to areas around stream banks under rock crevices, and lowland rainforests of southwestern Sri Lanka at 60–460m elevation, in Sinharaja Forest Reserve, Kanneliya Forest Reserve and Gilimale Forest Reserve. Also found in tea and rubber plantations, and home gardens of adjoining rainforests. However, it is not as common around human habitations as *D. melanostictus*.

Status Endemic.

IUCN Red List Category Critically Endangered.

Fig. 1 Lateral aspect

Fig. 2 Dorsal aspect

Fig. 3 Close-up of head indicating parotid gland (top arrow) and tympanum (bottom arrow)

Ferguson's Toad *Duttaphrynus scaber*

(Fergusongē gembā)

First described as *Bufo scaber* by J. G. Schneider in 1799. Redescribed as *B. fergusonii* by George Boulenger in 1892, and as *B. stomaticus* by Narayanan Rao in 1920. Currently known as *Duttaphrynus scaber. D. atukoralei* is now placed in the synonymy of *D. scaber.*

Size SVL 21–36mm.

Identification Features Body short and stout, limbs short and fingers have rough edges. Snout obtusely pointed and head broader than long. Nostrils nearer to tip of snout than to orbit. Canthal ridge angular. Canthal, preorbital, supraorbital, postorbital, supratympanic and parietal ridges also present. Loreal region concave. Dorsum has numerous small spinous tubercles. Smaller tubercles on venter and sides of head. Toes webbed and fingers not webbed. Tips of digits rounded. Can be distinguished from other *Duttaphrynus* species by presence of more or less oval and swollen parotid glands, and clusters of warts on dorsal side of body (Figs 1–3).

Colour Dorsal surface a mixture of light brown and grey with dark patches (Figs 1 & 2). In some individuals, two lateral bands (yellowish with tinge of red) on either side of body below base of parotid to groin. Throat, chest and belly yellowish-cream without blackish markings.

Habits Nocturnal terrestrial species. Like most amphibians, becomes active during the monsoon rains. By day hides in leaf litter, underneath rubble and decaying logs.

Habitat and Distribution Found in tropical dry forests, dry scrubland, grassland, coastal marshes, home gardens and rural farmland. Common around tanks and other aquatic habitats of lowland areas of dry zone. Widespread in lowland dry zone up to 300m above sea level, and coastal areas in wet zone. Recorded from Arachchikattuwa, Dambulla, Giritale, Kandalama, Mihintale, Puttalam, Trincomalee, Wilpattu, Galle, Atthidiya, Yala, Buttala and Batticaloa.

Status Native. Extralimital: India.

IUCN Red List Category Least Concern.

Fig. 1 Dorsolateral aspect

Fig. 2 Lateral aspect

Fig. 3 Oval and swollen parotid glands indicated by arrow

Family Dicroglossidae (Fork-tongued Frogs)
Frogs of the family Dicroglossidae were previously placed in the subfamily Dicroglossinae under the family Ranidae, but recent molecular phylogenetic analyses have shown that they deserve to be in a separate family. As the family consists of a diverse array of taxa, there are very few external features that unite them apart from the presence of a forked tongue. They have a free-living tadpole stage that possesses keratinous mouthparts. Approximately 213 species in 14 genera are known from Africa, the Middle East, South Asia and Southeast Asia. Thirteen species in five genera (*Euphlyctis, Hoplobatrachus, Minervarya, Nannophrys* and *Sphaerotheca*) have been recorded in Sri Lanka. These can be distinguished externally from other frogs in Sri Lanka by the presence of moist skin and absence of expanded fingertips and transverse skin folds on the dorsum. Most Sri Lankan species are associated with aquatic environments, except those in the genus *Sphaerotheca*. The genus *Nannophrys* is endemic to Sri Lanka.

Genus *Euphlyctis*
Euphlyctis are medium to large (40–70mm), entirely aquatic frogs. They can be differentiated from all other Sri Lankan dicroglossids by the presence of fully webbed toes and absence of longitudinal ridges on the back. Two species are known from Sri Lanka.

INDIAN SKIPPER FROG *Euphlyctis cyanophlyctis*

(Utpãtana mädiyã)

First described as *Rana cyanophlyctis* by J. G. Schneider in 1799. After several changes, (that is, to *Bufo, Dicroglossus* and *Occidozyga*), the current name is *Euphlyctis cyanophlyctis*. A recent study suggests that the Sri Lankan and South Indian populations merit the name *E. mudigere*.

Size SVL 35–65mm.

Identification Features Small to medium-sized frog with bulging eyes that are pointed upwards (Fig. 1). Snout obtusely pointed and head longer than broad. Canthal ridge indistinct. Loreal region concave. Dorsal side covered with granular tubercles. Belly smooth. Both limbs have numerous rounded or granular tubercles, while all fingers and toes are velvety. Toes fully and broadly webbed. Tips of digits pointed. Resembles *E. hexadactylus* and can be distinguished by its smooth throat (vs granular throat in *E. hexadactylus*) (Fig. 3).

Colour Dorsally brownish or olive-green with diffused brown spots, or dull green with dark brown spots (Figs 1 & 2). Limbs have dark brown cross-bands, which do not form complete cross-bands. Belly pearly-white. Throat has dark speckles (Fig. 3). Ventral warts black or white.

Habits Well adapted to aquatic life and active by both day and night. Prefers still water and often seen in water floating with its bulging eyes and nostrils above the water's surface, or at edges (Fig. 2). When disturbed, quickly jumps into the water or skips on the water's surface. Skipper frogs often climb on to the exposed parts of resting buffalos and cows in water to feed on flying insects on the mammals' bodies.

Habitat and Distribution Sometimes large numbers congregate in all types of water bodies (streams, ponds, isolated rock pools in degraded forests and human habitations). Occasionally found inside domestic wells. Widely distributed in Sri Lanka and found in all climatic zones up to 1,700m.

Status Native. Extralimital: India, Pakistan, Nepal, Bhutan, Bangladesh, Afganistan, Iran, Thailand and Myanmar.

IUCN Red List Category Least Concern.

Fig. 1 Dorsolateral aspect

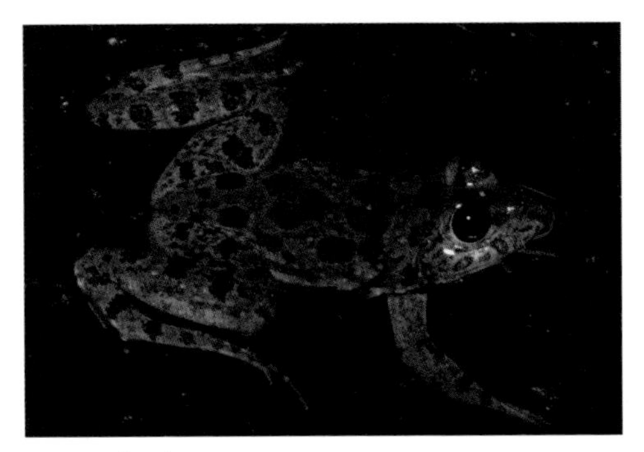

Fig. 2 Dorsolateral aspect

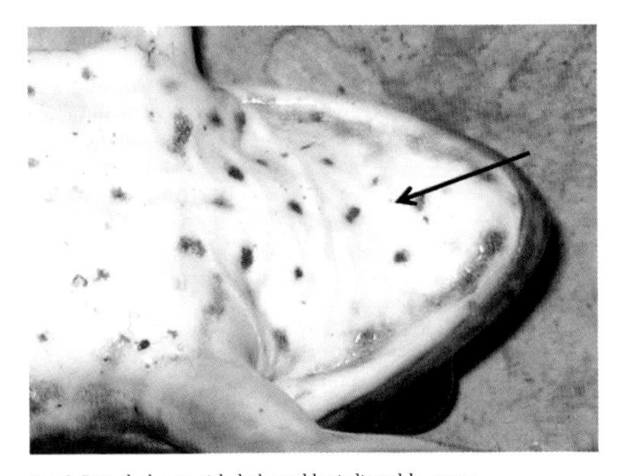

Fig. 3 Smooth throat with dark speckles indicated by arrow

SIX-TOE GREEN FROG/GREEN POND FROG *Euphlyctis hexadactylus*

(Sayäňgili palã mädiyã)

First described as *Rana hexadactyla* by R. P. Lesson in 1834. After several name changes (to *R. cutipora, R. saparoua, Phrynoderma cutiporum* and *Dicroglossus hexadactyla*), it is currently known as *Euphlyctis hexadactylus*.

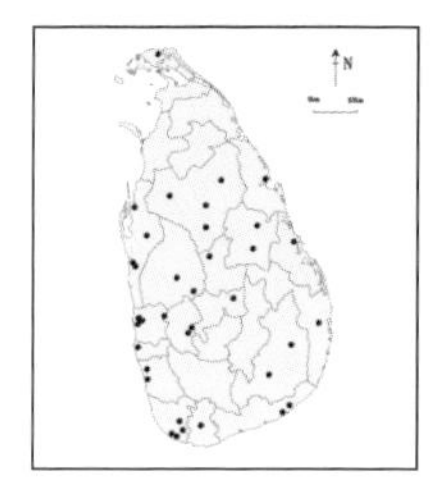

Size SVL 58–100mm.

Identification Features Large frog with smooth, moist skin and fully webbed feet. Head longer than broad. Canthal ridge indistinct. Loreal region concave. Large tubercles behind tympanum. Tips of digits pointed. Fingers not webbed. Nuptial pads of males are on base and inner side of first finger. Can be distinguished from *E. cyanophlyctis* by its granular throat (vs smooth throat in *E. cyanophlyctis*) (Fig. 3).

Colour Dorsal surface either bright green or a combination of brown and green (Figs 1 & 2). Broad, dark blotch along middle of back in some individuals. Belly pearly-white with tint of yellow (Fig. 3). Ventral aspects of thighs and legs have greyish-brown markings. Yellowish-white dorsolateral band from behind tympanum to groin. Some individuals may have dark patches in venter.

Habits Completely aquatic species that prefers ponds, rivers and lakes with aquatic vegetation. Occasionally found in domestic wells. Nocturnal species actively foraging at night, sometimes even on land.

Habitat and Distribution Highly abundant in lowland still-water bodies. Often found in sympatry with *Hoplobatrachus crassus*. Common and widely distributed in all climatic zones in Sri Lanka up to about 800m.

Status Native. Extralimital: India, Pakistan and Bangladesh.

IUCN Red List Category Least Concern.

Fig. 1 Dorsolateral aspect

Fig. 2 Dorsal aspect

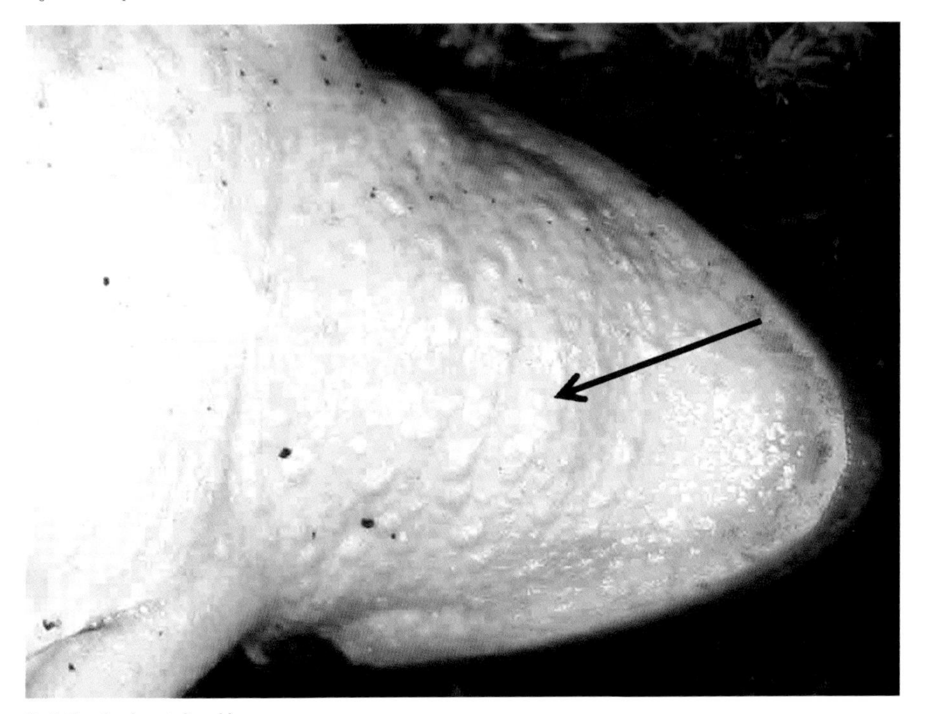

Fig. 3 Granular throat indicated by arrow

Genus *Hoplobatrachus*

Hoplobatrachus are large frogs that usually attain a SVL of at least 60mm – hence they are the largest frogs in Sri Lanka. They differ from all other Sri Lankan dicroglossids by the presence of blunt digit-tips, fully webbed feet and prominent longitudinal ridges on the skin of the back. Tadpoles of *Hoplobatrachus* are carnivores and often prey on other tadpoles. Two species are recorded in Sri Lanka.

Jerdon's Bull Frog *Hoplobatrachus crassus*

(Jerdongẽ hala mädiyā)

First described as *Rana crassa* by T. C. Jerdon in 1853. After revisions of the generic name (that is, to *Dicroglossus* and *Limnonectes*), the currently accepted name is *Hoplobatrachus crassus*.

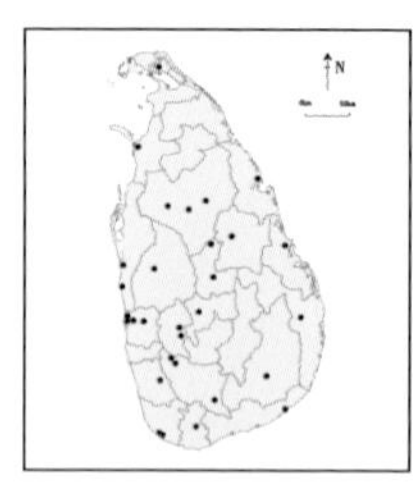

Size SVL 60–120mm.

Identification Features One of the largest frogs in Sri Lanka. Characterized by blunt digit-tips, fully webbed feet and prominent longitudinal ridges on skin of back (Figs 1 & 2). Head longer than broad. Supratympanic fold prominent. Tympanum distinct and rounded or oval in shape. Loreal region oblique. Well-developed cutaneous fringes on both sides of fingers, and well-developed cutaneous fold on outer side of fifth toe. Numerous white-tipped tubercles around vent and both limbs. Differs from *H. tigerinus*, the species it closely resembles in Sri Lanka, by relatively shorter hindlimbs (vs relatively longer hindlimbs in *H. crassus*) and shovel-shaped inner metatarsal tubercle (IMT) (vs elongated IMT in *H. tigerinus*) (Fig. 3).

Colour Yellowish-green dorsum with olive-brown or mixture of brown, yellow and green, with characteristic large, irregular, blackish or dark brown markings (Figs 1 & 2). Pearly-white belly. Black and white bars on lower jaws. Sides of body and ventral sides of thighs marbled with black and yellow (Fig. 2).

Fig. 1 Dorsolateral aspect

Habits Semi-aquatic species that prefers to live close to water. During dry season usually hides inside cracks in earth, in dried ponds and holes in stream banks, underneath logs and stones, and in various water bodies. Has even been seen inhabiting highly polluted drain water. Mostly active during rainy season.

Habitat and Distribution Mainly an aquatic species found in all shallow waters, reservoirs, rivers, marshland, paddy fields, drains and agro wells. Common and widely distributed in all climatic zones from sea level up to about 600m elevation.

Status Native species. Extralimital: India, Nepal and Bangladesh.

IUCN Red List Category Least Concern.

Fig. 2 Lateral aspect

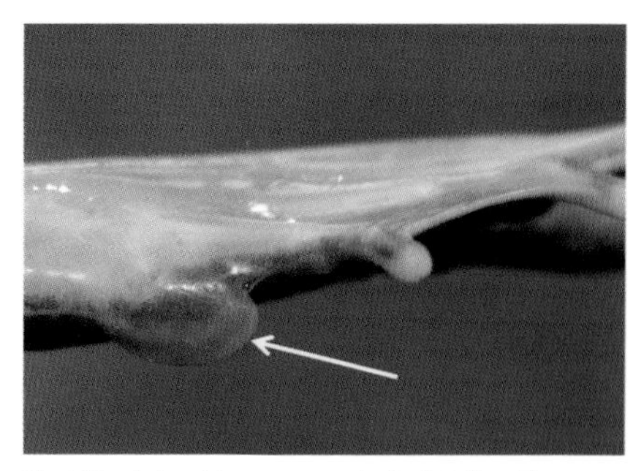

Fig. 3 Shovel-shaped inner metatarsal tubercle indicated by arrow

Indian Bull Frog *Hoplobatrachus tigerinus*

(Indiyanu hela mädiyã)

First described as *Rana tigerina* by F. M. Daudin in 1802. After several changes of the generic name (that is, to *Dicroglossus, Euphlyctis, Fejervarya* and *Limnonectes*), the current name is *Hoplobatrachus tigerinus.*

Size SVL 60–125mm.

Identification Features One of the largest frogs in Sri Lanka. Characterized by blunt digit-tips, fully webbed feet and prominent longitudinal ridges on skin of back. Head longer than broad. Supratympanic fold prominent. Tympanum distinct and rounded. Loreal region oblique. Interorbital distance nearly half the width of upper eyelid. Cutaneous fringes along outer toes. Differs from *H. crassus*, the species it closely resembles in Sri Lanka, by relatively longer hindlimbs (vs relatively shorter hindlimbs) and elongated inner metatarsal tubercle (vs shovel-shaped one in *H. crassus*).

Colour Dorsum brownish or mixture of brown and yellow, with characteristic large, irregular, blackish or dark brown markings. Belly white. Lips have black and white bars. Black and yellow marbles on sides of body and ventral sides of thighs. Limbs have black cross-bars. Black and white speckles on inner sides of thighs.

Habits Semi-aquatic species that prefers to live on land close to water bodies.

Habitat and Distribution The presence of this species in Sri Lanka is doubted as no individuals have been seen in the recent past.

Status Native (doubtful).

IUCN Red List Category Least Concern.

Genus *Minervarya*

Minervarya is a genus of relatively small frogs that grow to up to a maximum of 45mm. They can be distinguished from other dicroglossid genera by the incomplete webbing on the feet, blunt finger- and toe-tips, and the webbing on the fourth toe between the penultimate subarticular tubercle and tip. Three species are known from Sri Lanka, of which two are endemic. All Sri Lankan species were formerly placed in the genus *Fejervarya*. However, recent molecular phylogenetic analyses suggest that they should be placed in the genus *Minervarya*. Members of the genus *Fejervarya* are currently considered to be restricted in distribution to Southeast Asia and northeastern India.

Common Paddy Field Frog *Minervarya agricola*

(Sulaba hala mädiyã)

First described as *Rana agricola* by Jerdon in 1853. However, for many decades this South Asian taxon was erroneously referred to as *Limnonectes limnocharis*, a species restricted to Southeast Asia. After several name changes (some of the synonyms include *Limnonectes limnocharis, Fejervarya limnocharis, Zakerana limnocharis* and *Minervarya granosa*), the currently accepted name is *Minervarya agricola*.

Size SVL 19–45mm.

Identification Features Small frog with obtusely pointed snout, concave loreal region, partially webbed toes and distinct tympanum. Supratympanic fold extends from behind eye to bases of forelimbs. (Figs 1 & 2). Fingers and toes have dermal fringe on both sides. Thighs lightly granular. Can be distinguished from other *Minervarya* species in Sri Lanka by interrupted dorsal ridges on dorsum (vs uninterrupted dorsal ridges in *M. kirtisinghei* and *M. greenii*) (Fig. 3).

Colour Dorsally greenish, yellowish-brown

Fig. 1 Dorsolateral aspect

or olive-brown, with or without distinct yellow or cream-coloured vertebral line on dorsal side. Dark brown, 'W'-shaped marking extends from interorbital region along back, up to vent. Lips have 4–5 vertical bars. Backs of thighs marbled with black. Limbs contain dark brown cross-bands dorsally (Figs 1–3). Throat greyish-black or dark brown. Belly white or pale yellowish-white.

Habits Mainly an aquatic species, but frequents land far away from water even in urban areas. Hides under stones and logs during dry season. Usually floats on water, exposing its heads above the water, and dives quickly into the water when disturbed

Habitat and Distribution Commonly found in paddy fields, streams and other shallow aquatic habitats. One of the most common and most widely distributed frogs in all climatic zones below 1,400m.

Status Native. Extralimital: India.

IUCN Red List Category Least Concern.

Fig. 2 Dorsolateral aspect

Fig. 3 Dorsolateral aspect, with arrow indicating interrupted dorsal ridges

MONTANE PADDY FIELD FROG *Minervarya greenii*

(Lanka vel mädiyã)

First described as *Rana greenii* by G.A. Boulenger in 1905. After several generic name changes (that is, to *Euphlyctis*, *Limnonectes*, *Zakerana* and *Fejervarya*), the current name is *Minervarya greenii*.

Size SVL 30–45mm.

Identification Features Small to medium-sized frog that can be distinguished from *M. agricola* by uninterrupted dorsal ridges on dorsum (Figs 1 & 2). Snout pointed obtusely and head long rather than broad. Nostrils nearly halfway between tip of snout and eye. Dorsal sides of body have 6–8 prominent uninterrupted longitudinal ridges, and 3–4 mid-dorsal ridges that are usually not broken (Figs 1 & 2). Finger- and toe-tips blunt. Hindlimbs have fine warts. Distinguishable from *M. kirtisinghei*, the species it most closely resembles, by well-developed cutaneous fringe on inner edge of the first toe (absent in *M. kirtisinghei*) (Fig. 3). Supratympanic fold slightly prominent.

Colour Dorsum a mixture of brown and olive-green, or grey and olive-green with dark spots (Figs 1 & 2). Distinct yellow band on dorsal side. Lips have distinct black and white bars. Dorsal sides of limbs light olive with distinct cross-bars or blotches. Hindlimbs contain fine warts. Belly light yellowish-white. Ventral sides of thighs and forelimbs reddish-brown.

Habits Semi-aquatic species, commonly found in vegetation around pools, streams, waterholes and other water sources.

Habitat and Distribution Common at 1,700–2,200m above sea level in Central Highlands of Sri Lanka. Has been recorded in pools, streams and immediate surrounding vegetation at Horton Plains, Hakgala, Ohiya, Pattipola, Ambewela, Blackpool and Nuwara Eliya.

Status Endemic.

IUCN Red List Category Endangered.

Fig. 1 Dorsolateral aspect

Fig. 2 Lateral aspect

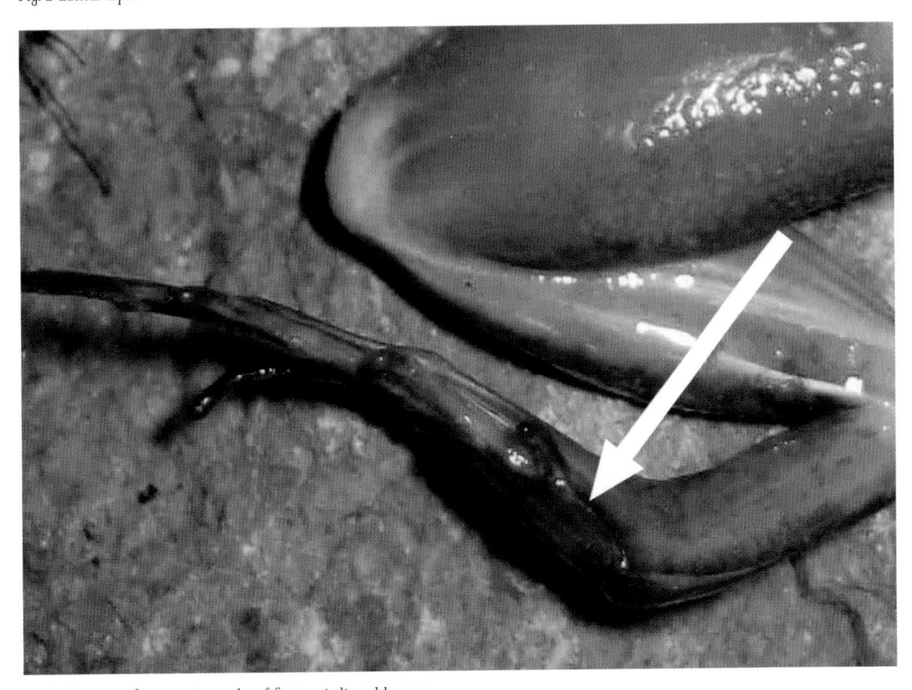

Fig. 3 Cutaneous fringe on inner edge of first toe indicated by arrow

Sri Lanka Paddy Field Frog *Minervarya kirtisinghei*
(Sri Lanka vel mädiyā̃)

First described as *Limnonectes kirtisinghei* by Kelum Manamendra-Arachchi and Dinesh Gabadage in 1996. After changes to the generic name (that is, to *Fejervarya* and *Zakerana*), the currently accepted name is *Minervarya kirtisinghei*.

Size SVL 25–44mm.

Identification Features Small to medium-sized frog that can be distinguished from *M. agricola* by uninterrupted dorsal ridges on dorsum (Figs 1 & 2). Loreal region concave. Supratympanic fold fleshy and prominent. Finger- and toe-tips blunt. Hindlimbs contain fine warts. Head longer rather than broad. Distinguished from *M. greenii*, the species it most closely resembles, by absence of well-developed cutaneous fringe on inner edge of first toe (present in *M. greenii*) (Fig. 3). Snout appears smoothly rounded when viewed laterally (vs pointed obtusely in *M. greenii*). Canthal ridge indistinct and smooth.

Colour Dorsal a combination of dark brown and grey with dark patches (Figs 1 & 2). Distinct yellow vertebral line on dorsal side in some individuals. Belly light brown with trace of yellow.

Habits Mainly an aquatic species, inhabiting streams, paddy fields and other aquatic habitats. Reported from places far away from water sources. May be encountered in drains near human habitats.

Habitat and Distribution Common species widely distributed in wet climatic zone, mainly below 1,600m. Common in the Central Highlands, Knuckles Mountain Range and Rakwana, and a few locations in the intermediate zone.

Status Endemic.

IUCN Red List Category Near Threatened.

Fig. 1 Dorsolateral aspect

Fig. 2 Dorsal aspect, with uninterrupted dorsal ridges indicated by arrow

Fig. 3 Ventral aspect of foot

Genus *Nannophrys*
Nannophrys are medium-sized frogs that usually grow to a maximum of 60mm. They can be distinguished from the rest of the Sri Lankan dicroglossids by the presence of dorsoventrally flattened bodies, horizontal pupils, blunt fingertips and partially webbed toes. They are always found on wet rock surfaces or rock crevices close to streams. Their flattened bodies allow them to live easily inside crevices or under boulders. They have semi-terrestrial tadpoles that live on wet rock surfaces. Mating frogs attach their eggs under wet rock surfaces, and the eggs hatch into a more developed tadpole stage than tadpoles of other frogs. The male frogs are known to guard the eggs until they hatch. The genus is endemic to the island. Four species are known in Sri Lanka.

Sri Lanka Rock Frog *Nannophrys ceylonensis*
(Lanka galpara mäḍiya)

First described as *Nannophrys ceylonensis* by A. C. L. G. Günther in 1868. Later identified as *Trachucephalus ceylanicus* by W. Ferguson in 1875, and the current name is *Nannophrys ceylonensis.*

Size SVL 33–53mm.

Identification Features Small to medium-sized frog with dorsoventrally flattened body, horizontal pupils, blunt fingertips and partially webbed toes (Figs 1 & 2). Head depressed and broader than long. Canthal ridge and loreal region concave. Head and trunk separated by narrow groove running transversely behind posterior level of eye. Dorsal surface has many granular warts. Sides of body and hind sides of forelimbs have prominent coarse tubercles. Can be distinguished from the rest of the *Nannophrys* species in Sri Lanka by its truncated snout when viewed laterally (Fig. 3).

Colour Dorsum a mixture of olive-green

Fig. 1 Dorsal aspect

and yellow marbled with brownish patches (Figs 1 & 2). Hindlimbs have distinct dark brown cross-bars. Belly white and chin and ventral sides of limbs diffused with light brown freckles.

Habits The flat body shape of this frog helps it to creep into narrow cracks in rocks. Tadpoles are semi-terrestial and capable of living on wet rock surfaces. Adults are habitat specialists that prefer wet rock surfaces. By day they can be seen on wet rock surfaces, within narrow cracks on rocks, and also under boulders along streams and cascades, in forests and even outside forests.

Habitat and Distribution Restricted to lowland wet zone of Sri Lanka up to 1,100m above sea level. Reported localities are: Avissawella, Ambuluwawa, Ambagamuwa, Bathalegala, Dolosbage, Dombagaskanda, Labugama, Lankagama Yagirala, Kanneliya, Kudawa, Kuruwita and Kithulgala.

Status Endemic.

IUCN Red List Category 2019 Vulnerable.

Fig. 2 Dorsolateral aspect

Fig. 3 Truncated snout indicated by arrow

Guenther's Rock Frog *Nannophrys guentheri*

(Guntargē galpara mädiya)

First described by G. A. Boulenger in 1882. Currently known only from two syntypes in the Natural History Museum, London, bearing the number BMNH 1947.2.5.20-21. No individuals matching its description have been seen in Sri Lanka for the last 80 years, and thus the species is currently considered to be extinct.

Status Endemic.

IUCN Red List Category Extinct.

Kirtisinghe's Rock Frog *Nannophrys marmorata*

(Kirtisinghage galpara mäḍiyä).

First described as *Nannophrys marmorata* by Parakrama Kirtisinghe in 1946. Was assigned as a subspecies of *Nannophrys ceylonensis* (*Nannophrys ceylonensis marmorata*), but was re-elevated to species status.

Size SVL 25–52mm.

Identification Features Small to medium-sized frog with dorsoventrally flattened body, horizontal pupils, blunt fingertips and partially webbed toes. Snout smoothly rounded. Head broader than long and depressed. Canthal ridge rounded and loreal region flat. Dorsum has scattered, fine, white-tipped tubercles (Figs 1 & 2). Venter smooth. Can be distinguished from N. *ceylonensis* by rounded snout when viewed laterally. Distinguished from N. *guentheri* by rounded snout when viewed laterally and white-tipped tubercles on upper surface of head (Fig. 3). Distinguished from N. *naeyakai* by closely placed distal subarticular and penultimate subarticular tubercles on fourth toe (vs well-separated distal subarticular and penultimate subarticular tubercles in N. *naeyakai*) (Fig. 4).

Colour Dorsum of body and limbs brown or yellowish-brown marbled with brownish patches (Figs 1 & 2). Irregular patches on dorsal sides of limbs. Dorsally a yellowish tinge with a trace of grey. Belly and ventral sides of limbs white.

Habits Nocturnal species with occasional diurnal activities. Mainly hides in narrow cracks in rocks by day. Flat body shape helps it to creep into narrow cracks in rocks. Tadpoles semi-terrestrial and may be seen on wet rock surfaces and even underneath

Fig. 1 Dorsolateral aspect of male

small rocks on hot sunny and windy days.

Habitat and Distribution Occurs on rocks and under stones on wet, flat and rocky surfaces or logs along streams and cascades. Mainly confined to the Knuckles Mountain Range, and may be seen in Gammmaduwa, Emadawa, Laggala, Lakegala, Kahatagolla, Pitawala Pathana, Rathna Ella and Ududumbara.

IUCN Red List Category: Endangered.

Fig. 2 Dorsolateral aspect of female

Fig. 3 Lateral aspect of head, with rounded snout indicated by arrow

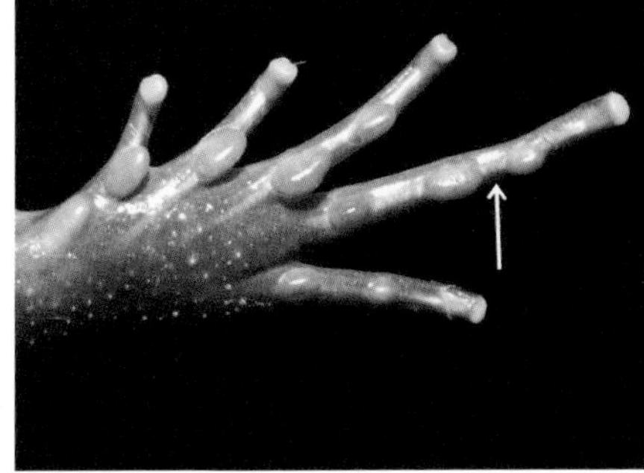

Fig. 4 Ventral aspect of foot, closely placed subarticular tubercles on fourth toe indicated by arrow

SRI LANKA TRIBAL ROCK FROG *Nannophrys naeyakai*

(Nae-yak galpara mädiyã)

First described as *Nannophrys naeyakai* by Suranjan Fernando, Mendis Wickramasighe and Roshan Rodrigo in 2007. However, first illustrated and reported in 2004 as an unknown species of *Nannophrys* from Nilgala by Anslem de Silva and colleagues in 2004.

Size SVL 24–44mm.

Identification Features Small to medium-sized frog with dorsoventrally flattened body, horizontal pupils, blunt fingertips and partially webbed toes (Figs 1 & 2). Nostrils closer to tip of snout than to eye and dorsally flattened (not elevated from head). Loreal region concave. Rounded tympanum with horizontal diameter equivalent to half of orbit diameter. Tubercles on dorsal portions of head, shoulders and legs in scattered pattern. Can be differentiated from *N. marmorata*, the species it most closely resembles, by well-separated distal subarticular and penultimate subarticular tubercles on fourth toe (vs closely placed distal subarticular and penultimate subarticular tubercles in *N. marmorata*) and dense yellow stripes on hindlegs (vs marbled pattern in *N. marmorata*) (Fig. 3).

Colour Dark brown body colour with dense yellow spots (Figs 1 & 2). Yellow thick stripes runs across legs perpendicularly to body (one is along upper and lower portions of forearms, and three stripes across hindlegs). Ventral side yellow without any markings.

Habits Nocturnal species. Lives on wet rock surfaces along seasonal streams, and active for a short period of time during the year from December to February. During drought retreats into cracks in rocks.

Habitat and Distribution Only known from four localities in Ampara and Monaragala Districts, including Nilgala, Kokagala, Padiyatalawa and Pitakumbura at 200–620m above sea level.

Status Endemic.

IUCN Red List Category Endangered.

Fig. 1 Dorsolateral aspect

Fig. 2 Dorsal aspect

Fig. 3 Ventral aspect

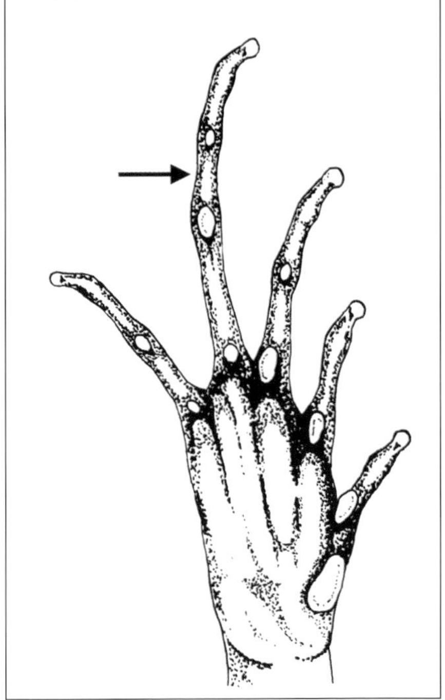

Fig. 4 Ventral aspect of foot, with well-separated subarticular tubercles on fourth toe indicated by arrow

Genus *Sphaerotheca*
Sphaerotheca are medium-sized frogs that grow to a SVL of up to 60mm. They are characterized by their globular-shaped bodies and short limbs. They can be distinguished from other Sri Lankan dicroglossids by their blunt digit-tips, partially webbed toes, fourth toe webbing to between the penultimate subarticular tubercle and antepenultimate subarticular tubercle, and an elongated, compressed, shovel-shaped inner metatarsal tubercle on the feet. They are burrowing species, and the shovel-shaped inner metatarsal tubercles on the feet aid in burrowing. Two species are known from Sri Lanka.

JERDON'S SAND FROG *Sphaerotheca pluvialis*
(Tuniri väli mädiya)

First described as *Pyxycephalus pluvialis* by T. C. Jerdon in 1853. However, the population in Sri Lanka was initially assigned to *Sphaerotheca breviceps*. The species also went through generic name changes (that is, to *Pyxicephalus* and *Tomopterna*). Its current name is *Sphaerotheca pluvialis*.

Size SVL 33–53mm.

Identification Features Small to medium-sized, stocky frog with short snout, globular-shaped body, short limbs, blunt digit-tips, partially webbed toes and elongated, compressed, shovel-shaped inner metatarsal tubercle on feet. Head small and broader than long. Snout rounded dorsally and truncated laterally. Canthal ridge rounded and loreal region concave. Tympanum distinct and vertically oval. Supratympanic fold distinct and extends from backs of eyes to bases of forelimbs. Smooth or granular dorsum with small, rounded or elongated warts (Figs 1 & 2). Ventral side granular. Throat and chest smooth. Thin dermal fringes on both sides of second, third and forth toes, and inner side of fifth toe. *Sphaerotheca pluvialis* and *S. rolandae* are sympatric species and look similar. *S. pluvialis* can be differentiated from *S. rolandae* by absence of tubercle at the tibiotarsal

Fig. 1 Lateral aspect

articulation (present in *S. rolandae*), and distal subarticular tubercle on first toe (absent in *S. rolandae*) (Fig. 3).

Colour Light to dark mud-brown dorsum. Three light brown or cream-coloured lines run on either side of body and one along mid-dorsum. Dark brown cross-bars on dorsal sides of limbs. Ventral aspect light coloured.

Habits Terrestrial and nocturnal species usually active during the monsoon rain. By day buries itself in loose sand or hides underneath rocks or logs. Burrows into the soil with the special 'shovel-like' modified tubercle on its hindlegs.

Habitat and Distribution Occurs in home gardens, grassland and forests. Widely distributed and common in lowlands of intermediate and dry zones, including Andigama, Kandalama, Galnewa, Dambulla, Kurunegala, Mihintale, Polonnaruwa, Sevanagala, Trincomalee, Udawalawe, Mannar and Yala.

Status Native. Extralimital: India.

IUCN Red List Category Least Concern.

Fig. 2 Dorsolateral aspect

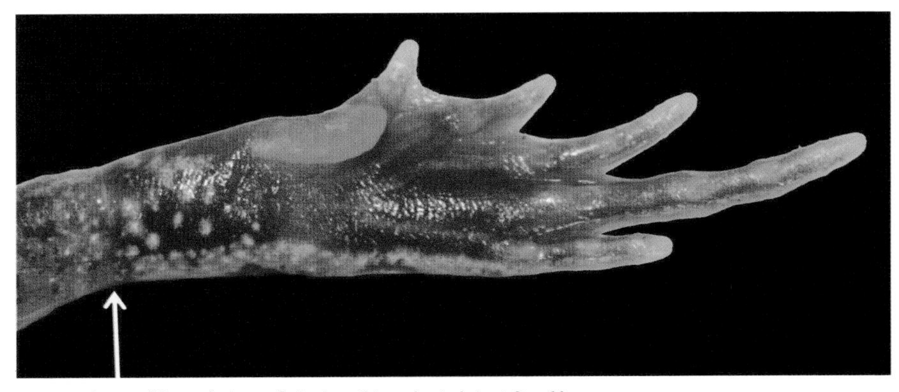

Fig. 3 Ventral aspect of foot, with absence of tubercle at tibiotarsal articulation indicated by arrow

MARBLED SAND FROG *Sphaerotheca rolandae*

(Lapavan väli mädiyã)

First described as *Rana breviceps rolandae* by Alan Dubois in 1983 from Kurunegala. After a number of changes to the genus (including to *Tomopterna*), the currently accepted name is *Sphaerotheca rolandae*.

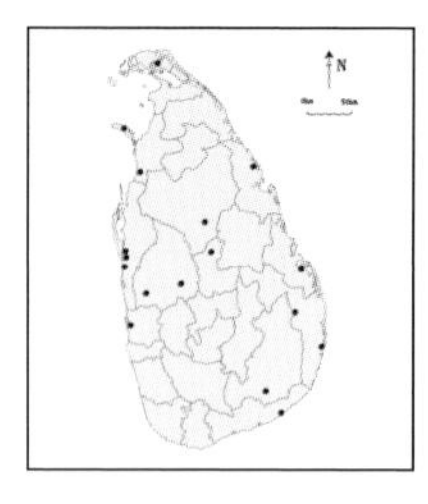

Size SVL 32–45mm.

Identification Features Small to medium-sized, stocky frog with short snout, globular-shaped body, short limbs, blunt digit-tips, partially webbed toes and elongated, compressed, shovel-shaped inner metatarsal tubercle on feet. Snout rounded laterally. Head small, distinct and broader than long. Canthal ridge rounded and loreal region concave. (Figs 1 & 2). Tympanum distinct, rounded or vertically oval. Supratympanic fold extends from backs of eyes to bases of forelimbs. Dorsum smooth or granular with rounded or elongated scattered tubercles. Ventral side granular. Throat and breast smooth. Dermal fringes on sides of second, third and fourth toes and inner side of fifth toe. Can be differentiated from *S. pluvialis* by tubercle at the tibiotarsal articulation (absent in *S. pluvialis*) and absence of distal subarticular tubercle on first toe (present in *S. pluvialis*) (Fig. 3).

Colour Dorsum light brown or yellow and symmetrically marbled with dark brown spots (Figs 1 & 2). There may be a yellow vertebral line or band on the dorsum. Dark brown patches on dorsal sides of limbs. Black or dark brown interorbital bar. Ventral aspect light and granular.

Habits Terrestrial and nocturnal. By day, burrows into the soil by working with the special 'shovel-like' modified inner metatarsal tubercle. Active during monsoon period. Unique among other leaping frogs as it is capable of running.

Habitat and Distribution Occurs in home

Fig. 1 Dorsolateral aspect

gardens, grassland and forests. Widely distributed in both dry and intermediate zones up to 200m in elevation. Known locations include Andigama, Buttala, Kandalama, Mihintale, Kurunegala Puttalam, Meerigama and Ritigala.

Status Endemic.

IUCN Red List Category Least Concern.

Fig. 2 Lateral aspect

Fig. 3 Tubercle at tibiotarsal articulation of each foot indicated by arrows

Family Microhylidae (Narrow-mouth Frogs)
Microhylids are usually small to medium-sized frogs that grow to a maximum of 100mm. They have several body shapes, ranging from squat and small-headed globular, to toad-like, and to arboreal frogs with expanded digit-tips. Their pupils are horizontal or round, and they are characterized by a narrow head region with a teardrop-shaped, globose body, stout hindlegs and short snout. Reproduction can include direct and indirect development, and amplexus is axillary. The tadpoles lack keratinous beaks or denticles and hence are filter feeders. Worldwide, 693 species in 63 genera are known, and are categorized in 11 subfamilies. Microhylids are distributed throughout the world in warm temperate and tropical regions. Ten species in two genera (*Microhyla* and *Uperodon*) are known from Sri Lanka, seven of which are endemic.

Genus *Microhyla*
Microhyla are small frogs with flat, teardrop- or triangular-shaped bodies that do not usually exceed an SVL of 35mm. The skin is moist and either smooth or with scattered minute granules. The fingertips are blunt or form slightly expanded discs. Fifty species are known from South to Southeast Asia. Four are known from Sri Lanka, of which three are endemic.

Karunaratne's Narrow-mouth Frog *Microhyla karunaratnei*

(Karunaratnage muva patu mädiyã)

First described as *Microhyla karunaratnei* by Prithiviraj Fernando and Mahendra Siriwardhane in 1996.

Size SVL 15–21mm.

Identification Features Small frog (smallest *Microhyla* species in Sri Lanka) with flat, triangular-shaped body. Skin moist and smooth with a few scattered minute granules. Fingertips and toe-tips form slightly expanded round discs. Head broader than long. Snout rounded dorsally and laterally. Nostrils located dorsolaterally. Pupil of eye round. Canthal ridge rounded and loreal region oblique. Tympanum not distinct. Subarticular tubercles prominent and rounded or oval in shape. Toes webbed. (Figs 1 & 2). Toes have lateral fringes on both inner and outer sides. Thin median ridge from tip of snout to vent. Vent area and hind sides of thighs have tubercles. Differs from all other *Microhyla* species in Sri Lanka by median clefts (a groove seen from dorsal side) in toe-tips (absent in all other *Microhyla* species), and whitish belly with black or brown large blotches (all other *Microhyla* species in Sri Lanka have white bellies) (Fig. 3).

Colour Dorsum a mixture of grey and brown with dark greyish-brown, symmetrically arranged pattern that begins at mid-level of eyes and widens as it reaches posterior region of body. Dorsal sides of limbs have dark greyish-brown, incomplete cross-bars. Pinkish grey-brown lateral stripe extends from eye to groin. (Figs 1 & 2). Belly white in juveniles and patterned in adults (Fig. 3).

Habits Nocturnal species. By day hides under wet and moist leaf litter, decaying logs and rubble.

Habitat and Distribution Restricted to wet forests and degraded habitats, around anthropogenic habitats and agricultural fields in Rakwana Mountains (Morningside), Rajawaka and Balangoda at 500–1,100m in Sabaragamuwa province.

Status Endemic.

IUCN Red List Category Critically Endangered.

Fig. 1 Dorsolateral aspect

Fig. 2 Dorsolateral aspect

Fig. 3 Ventral aspect

Mihintale Red Narrow-mouth Frog *Microhyla mihintalei* (Mihintale rathu muvapatu mädiyã)

Formerly considered to be *M. rubra*, which was also shared with India. However, morphological, bioacoustic and genetic comparisons indicated that the Sri Lankan population of *M. rubra* was distinct from the Indian population and deserved its own name. The Sri Lankan population was designated as *M. mihintalei*.

Size SVL 21–28mm.

Identification Features Small frog with flat, triangular-shaped body. Skin moist and smooth, with a few scattered minute granules. Body small and slender. Canthal ridge rounded and loreal region round. Tympanum indistinct or obscured by layer of skin without clearly defined borders. Supratympanic fold distinct. Head small. (Figs 1 & 2). Ventral surface smooth. Dorsal parts of forelimbs, thighs, tibias and tarsi smooth to shagreened. Toes rudimentarily webbed. Subarticular tubercles prominent and oval shaped. Differs from *M. karunaratnei* and *M. zeylanica* by blunt digit-tips (vs slightly round, expanded digit-tips in *M. karunaratnei* and *M. zeylanica*). Differes from *M. ornata* by shovel-shaped metatarsal tubercles on feet (normal metatarsal tubercles in *M. ornata*).

Colour Dorsum light reddish-brown. Faint light greyish or brownish band extends from behind eye to groin region. Tympanic area and lateral sides of head dark greyish. Groin light grey with black patches. Two distinct narrow black streaks run from snout to groin. Thighs and tibias light brown with irregular black markings. Narrow black patch visible from anal opening to knee. Throat and buccal pouch dark blackish-brown. (Figs 1–3).

Habits Nocturnal species. Hides inside borrows in the ground during the day. Lays eggs as loosely arranged sheets on the water's surface, mainly in ephemeral pools. Tadpoles form loose shoals.

Habitat and Distribution Common species reported from shaded areas, close to stream and river banks, or fairly stable ephemeral pools during rainy season. Found in lowland dry and intermediate zones at up to 500m elevation.

Status Endemic.

IUCN Red List Category Least Concern.

Fig. 1 Dorsal aspect

Fig. 2 Lateral aspect

Fig. 3 Ventral aspect

ORNATE NARROW-MOUTH FROG *Microhyla ornata*

(Visituru muvapatu mädiyã)

First described as *Engystoma ornatum* by A. M. C. Duméril and G. Bibron in 1841. After several changes to the generic name (synonyms: *Engystoma ornatum* and *Siphneus ornatum*) and specific name (*Engystoma malabaricum, E. carnaticum, Diplopelma ornatum, D. carnaticum* and *Microhyla carnatica*), the current name is *Microhyla ornata*.

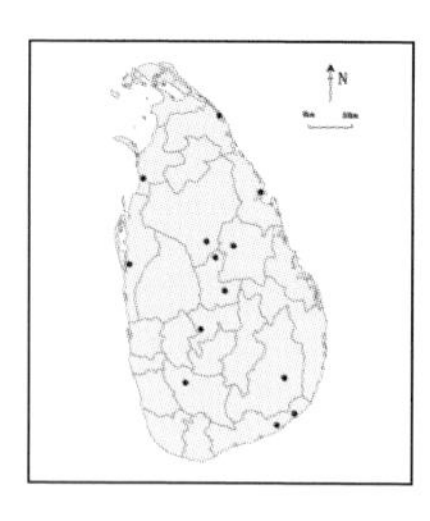

Size SVL 17–25mm.

Identification Features Small frog with flat, traingular-shaped body. Skin moist, smooth with a few scattered minute granules. Head broader than long. Snout pointed obtusely when viewed laterally. Nostrils located dorsolaterally. Canthal ridge rounded and loreal region concave. Tympanum not distinct. Supratympanic fold absent. Granular fold from posterior corners of eyes to groin (Figs 2 & 3). Hind sides of thighs granular and venter smooth. Toes rudimentarily webbed. Tips of fingers and toes rounded, without discs. Lateral fringes on toes. Subarticular tubercles on fingers are rounded. Outer edge of fifth toe has a cutaneous fringe. Differs from *M. karunaratnei* and *M. zeylanica* by blunt digit-tips (vs slightly rounded, expanded digit-tips in *M. karunaratnei* and *M. zeylanica*). Differs from *M. mihintalei* by normal (round) metatarsal tubercles on feet (shovel-shaped metatarsal tubercles in *M. mihintalei*).

Colour Dorsum light reddish-brown, greyish-brown or yellowish-brown with dark reddish-brown or brown, symmetrically arranged pattern that begins at mid-level of eyes and widens at posterior region of body (Figs 2 & 3). Dorsal sides of limbs have a few blotches. Belly whitish and throat dark (Fig. 1).

Fig. 1 Ventral aspect

Habits Nocturnal species that hides under leaf litter, decaying logs and rubble by day. Calling congregations can be seen during the monsoon rains, around water bodies such as ponds, agro-wells and paddy fields.

Habitat and Distribution Mainly restricted to anthropogenic habitats, including home gardens, paddy fields and plantations. Common species that mainly lives in lowlands of dry zone and a few locations in wet zone from sea level up to 500m above.

Status Native species. Extralimital: India, Nepal, Bangladesh, Bhutan and Pakistan.

IUCN Red List Category Least Concern.

Fig. 2 Dorsal aspect

Fig. 3 Lateral aspect

Sri Lanka Narrow-mouth Frog *Microhyla zeylanica*

(Lankã muvapatu mädiyã)

First described as *Microhyla zeylanica* by H.W. Parker and W. C. Osman-Hill in 1949.

Size SVL 24mm.

Identification Features Small frog with teardrop-shaped body. Skin moist and smooth with a few scattered minute granules. Head broader than long. Snout rounded when viewed dorsally and laterally. Nostrils arranged in middle with equal distance from tip of snout and anterior edge of orbit. Canthal ridge rounded. Loreal region convex and oblique. Tympanum not distinct and supratympanic fold present. Dorsum smooth or tuberculated. Raised median ridge from tip of snout to vent. Granular fold from posterior corner of eye to middle flank. Throat separated from pectoral region by transverse groove. (Figs 1 & 2). Venter smooth. Toes partially webbed. Tips of fingers rounded with poorly developed discs. Tips of toes rounded or triangular dilating into discs. Outer edge of fifth toe has prominent cutaneous fringe. Metatarsal tubercles oval shaped or elongated. Differs from *M. mihintalei* and *M. ornata* by slightly round, expanded digit-tips (vs blunt in *M. mihintalei* and *M. ornata*). Differs from *M. karunaratnei* by absence of median clefts in toe-tips (vs median clefts in toe-tips in *M. karunaratnei*) and whitish belly (vs whitish belly with black or brown large blotches in *M. karunaratnei*) (Fig. 3).

Colour Dorsum a mixture of reddish-brown and light brown with dark brown markings starting from back of eye and gradually extending up to groin (Figs 1 & 2). Loreal region and supratympanic fold black. Some individuals have a dark brown, symmetrically arranged pattern on dorsal side. White spots on lateral sides. Belly yellowish-white (Fig. 3), and limbs with or without cross-bars.

Fig. 1 Dorsolateral aspect

Habits Characteristic loud call. Tadpoles lie motionless beside *Eriocaulon* flowers, resembling the flowers, a strategy to avoid predators.

Habitat and Distribution Typically found around aquatic habitats at bases of grass tussocks (*Chrysopogon zeylanicus, Garnotia mutica*), and clumps of the dwarf bamboo *Sinarundinaria densifolia*. Mainly confined to high-altitude (>1,700m) montane grassland and forests (such as Horton Plains) of central Sri Lanka.

Status Endemic.

IUCN Red List Category Endangered.

Fig. 2 Lateral aspect

Fig. 3 Ventral aspect

Genus *Uperodon*
Uperodon species have globose or flat, teardrop-shaped bodies. Most possess triangular-shaped, sticky fingertips that aid them in climbing (except *Uperodon systoma*). Five species are known from this genus in Sri Lanka, with four being endemic.

Nagao's Pug Snout Frog *Uperodon nagaoi*

(Nagaoge moṭa hombu mädiya)

First described as *Ramanella nagaoi* by Kelum Manamendra-Arachchi and Rohan Pethiyagoda in 2001. Later it was placed in the genus *Uperodon*, and thus the current name is *Uperodon nagaoi*.

Size SVL 26–32mm.

Identification Features Small frog with flat, teardrop-shaped body and triangular-shaped fingertips. Skin moist with a few scattered tubercles on body. Nostrils closer to tip of snout than to eye and located dorsolaterally. Head broader than long. Canthal edges rounded and loreal region oblique and flat. Tympanum indistinct. (Figs 1 & 2). Superficially similar to *U. obscurus*, *U. rohani* and *U. palmatus*. Differs from *U. rohani* by dark brown belly (vs whitish belly in *U. rohani*). Differs from *U. obscurus* and *U. palmatus* by rudimentary webbing on toes (vs well-developed webbing in *U. obscurus* and *U. palmatus*).

Colour Dorsal side dark greyish-brown with brick-red patches (Fig. 1). Reddish-orange patches on snout, sides of dorsum, abdomen, around vent and on limbs. Distinct reddish-orange cross-bars on proximal end of tibias and tibiotarsal articulation. Belly dark purple and grey with white spots (Fig. 2).

Habits Eggs deposited in tree-holes, where tadpoles develop. Egg clutches with nearly 50 eggs are hung inside walls of tree-holes a few centimetres above the water level. Egg-laying tree-holes are situated about 30cm–9m above ground on large tree trunks (see p. 13, Fig. 16). Several females use a single tree-hole as the breeding site.

Fig. 1 Dorsolateral aspect

Ants are potential egg predators.

Habitat and Distribution Natural habitat is tropical moist lowland rainforest. Reported only from few localities in Sri Lanka, including Hiyare, Kanneliya, Kottawa, Kitulgala, Udamaliboda and Sinharaja.

Status Endemic.

IUCN Red List Category Vulnerable.

Fig. 2 *Ventral aspect*

Fig. 3 *Breeding habitat*

Brown Pug Snout Frog *Uperodon obscurus*

(Dumburu motahombu mädiyã)

First described as *Callula obscura* by A. C. L. G. Günther in 1864. Later recognized as *Ramanella obscura* by H.W. Parker in 1934. Since the genus *Ramanella* was synonymized with the genus *Uperodon*, the present name is *Uperodon obscurus*.

Size SVL 27–35mm.

Identification Features Small frog with flat, teardrop-shaped body and triangular-shaped fingertips. Skin moist with a few scattered tubercles on body. Head broader than long. Nostrils closer to tip of snout than to eyes. Canthal ridge rounded and loreal region vertical. Venter smooth. (Figs 1 & 2). Interorbital width greater than width of upper eyelid. Tympanum and supratympanic fold distinct. Fingers not webbed. Toe-tips without triangular dilations. Superficially similar to *U. nagaoi*, *U. rohani* and *U. palmatus*. Differs from *U. rohani* and *U. nagaoi* by well-developed webbing (vs rudimentary webbing in *U. rohani* and *U. nagaoi*). Differs from *U. palmatus* by webbing in fourth toe to penultimate subarticular tubercle or antepenultimate subarticular tubercle or between them on outer side (vs fourth-toe webbing to distal subarticular tubercle on outer side in *U. palmata*) (Fig. 3).

Colour Dorsal side dark greyish-brown with brick-red patches (Fig. 1). Black dorsal marking behind eyes that broadens in mid-dorsum and posteriorly. Limbs have black cross-bars. Belly a mixture of black or grey with white spots (Fig. 2).

Habits By day, adults hide under leaf litter, logs, crevices and similar. Produces a sticky, white, offensive-smelling substance from glands in body when handled. Breeds in natural nesting sites, that is, water-filled tree-holes (phytothelmata) and also drains. Egg clutch ranges from 50 to 250 eggs and usually floats on the water's surface. During the rainy season males congregate in temporary and permanent water sources and call to attract females.

Habitat and Distribution Commonly seen around anthropogenic habitats and also undisturbed forests. Mainly confined to

Fig. 1 Dorsolateral aspect

lowland wet zone up to 1,200m above sea level, and commonly seen around Kandy, Peradeniya, Gelioya, Gampola, Knuckles, Sinharaja and many other localities in lowland wet zone.

Status Endemic.

IUCN Red List Category Near Threatened.

Fig. 2 *Ventral aspect*

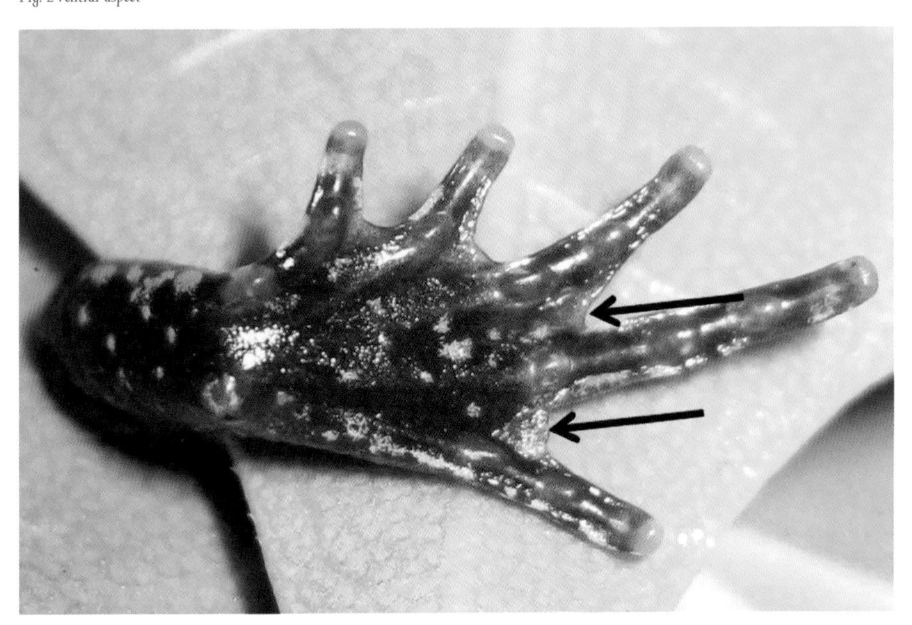

Fig. 3 *Ventral aspect of foot, with webbing indicated by arrows*

HALF-WEBBED PUG SNOUT FROG *Uperodon palmatus*

(Patala-pa motahombu mädiyã)

First described as *Ramanella palmata* by H.W. Parker in 1829. The genus *Ramanella* was later synonymized with *Uperodon*, and thus the current name is *Uperodon palmatus*.

Size SVL 27–35mm

Identification Features Small frog with flat, teardrop-shaped body and triangular-shaped fingertips. Skin moist with scattered tubercles on body. Head broader than long. Snout short and sub-equal to horizontal diameter of eye. Nostrils nearer to tip of snout than to anterior borders of eyes. Canthal ridge and tympanum indistinct and loreal region vertical. (Fig. 1). Venter smooth (Fig. 2). Toes contain lateral fringes. Two palmar tubercles, with inner one larger than outer one. Superficially similar to *U. obscurus*, *U. rohani* and *U. nagaoi*. Differs from *U. rohani* and *U. nagaoi* by webbing between toes (toe webbing rudimentary in *U. rohani* and *U. nagaoi*). Differs from *U. obscurus* by webbing in fourth toe to distal subarticular tubercle on outer side (vs fourth-toe webbing to penultimate subarticular tubercle or antepenultimate subarticular tubercle or between them on outer side in *U. obscurus*) (Fig. 3).

Colour Dorsum a mixture of orangish-brown and grey with dark brown patches (Fig. 1). Black patches laterally on both sides of dorsum and also patch around vent. Black patch on dorsum, which broadens between shoulders and sacral region. Loreal and tympanic regions black. Limbs have black patches or cross-bars. Belly a mixture of brown and grey with white spots (Fig. 2).

Habits Rare nocturnal species that rests under loose stones and rubble, inside decaying logs and underneath tree bark during the day. Known to inhabit tree-holes situated about 4–6m above ground level. Juveniles found on banks of ponds.

Habitat and Distribution Occurs in side forests, grassland and home gardens. Restricted to central Sri Lanka above

Fig. 1 Dorsolateral aspect

1,200m, including Ambewela, Blackpool, Hakgala, Horton Plains, Nuwara Eliya, Rilagala, Sri Pada and Pattipola.

Status Endemic.

IUCN Red List Category Endangered.

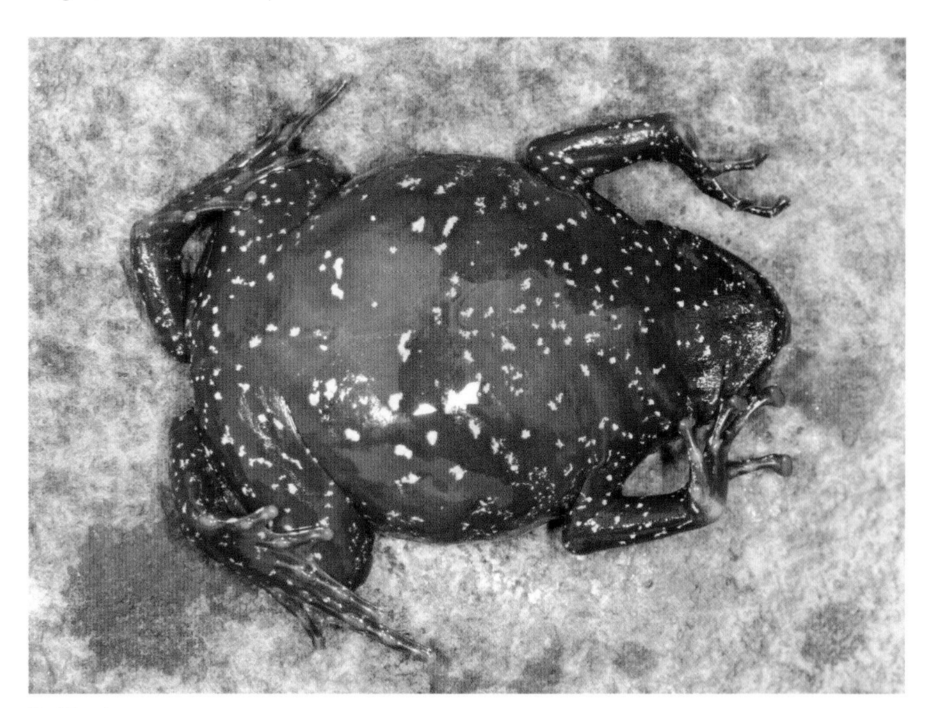

Fig. 2 Ventral aspect

Fig. 3 Ventral aspect of foot

Rohan's Pug Snout Frog *Uperodon rohani*

(Rohange motahombu mädiyã)

Formerly considered to be *Uperodon variegata* (formerly *Ramanella variegata*), which was also shared with India. However, morphological, bioacoustic and genetic comparisons indicated that the Sri Lankan population was distinct from the Indian populations and deserved its own name.

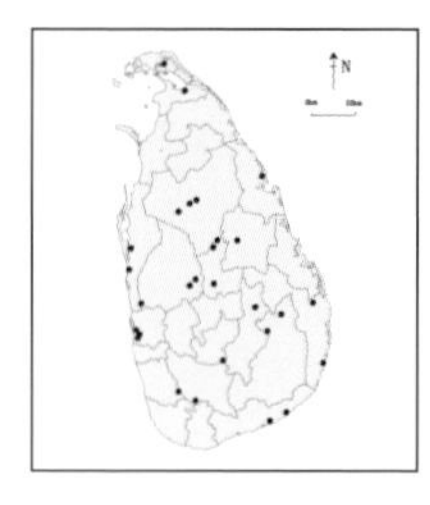

Size SVL 26–35mm.

Identification Features Small frog with flat, teardrop-shaped body and triangular-shaped fingertips. Skin moist and mostly smooth. Head small and less than one-fourth of body length. Loreal region rounded and canthal ridge round. Interorbital space wider than upper eyelid. Supratympanic fold distinct. Dorsum, snout and area between eyes shagreened. Sides of head, back of posterior part, and upper and lower flanks smooth; dorsal surfaces of forelimbs, thighs, tibias and tarsi smooth to shagreened. Ventral surfaces smooth. (Figs 1 & 2). Webbing on toes rudimentary. Superficially similar to *U. obscurus*, *U. palmatus* and *U. nagaoi*. It differs from all of them by presence of whitish belly (*U. obscurus*, *U. nagaoi* and *U. palmatus* have dark brown or black bellies) (Fig. 3).

Colour Dorsum yellowish to dark brown with pair of irregular olive-yellow longitudinal streaks extending from upper eyelids up to vent (Figs 1 & 2). Scattered olive-yellow spots and blotches especially towards dorsolateral surfaces and posterior part of back. Larger olive-yellow blotch in centre of dorsum. Lateral sides of head brown with scattered olive-yellow spots near margins of mouth. Anterior and posterior parts of groin light brown. Throat light brown with bluish-black calling patch and irregular light grey speckles (Fig. 3).

Habits Commonly found underneath dead bark of trees and under logs and stones.

Fig. 1 Dorsolateral aspect

During dry periods, can be seen in large numbers in bathrooms. Highly capable of moving up and down on walls and tree trunks. Eggs have been seen in temporary water puddles and roadside drains in November–January.

Habitat and Distribution Very common species that is easily seen in and around anthropogenic habitats, secondary forests and grassland. Widely distributed throughout in dry and intermediate zones of Sri Lanka up to an elevation of 350m above sea level.

Status Endemic.

IUCN Red List Category Least Concern.

Fig. 2 Dorsolateral aspect

Fig. 3 Ventral aspect

Balloon Frog *Uperodon systoma*
(Bälun mädiyã)

First described as *Rana systoma* by J. G.T. Schneider in 1799. Several different names were later used, including *Bombinator systoma, Engystoma marmoratum, Bufo marmoratus, Uperodon marmoratum, Hyperodon marmoratum, Cacopus systoma* and *Systoma marmoratum.* Currently accepted name is *Uperodon systoma.*

Size SVL 45–66mm.

Identification Features Medium-sized frog (largest microhylid in Sri Lanka) with globose, teardrop-shaped body. Skin moist with a few scattered tubercles. Head broader than long. Snout short. Nostrils nearer to eye than to tip of snout. Tympanum hidden and supratympanic fold present. Soft and soggy skin, and appears like a mass of loose jelly. Dorsum smooth or slightly tubercular. Snout, throat and limbs have small tubercles. (Figs 1 & 2). Venter smooth or granular, and anal region granular. Toes medially webbed. Two shovel-shaped metatarsal tubercles, with length of outer one subequal to half that of inner one. Thighs partially hidden. Differs from *U. taprobanicus*, the species it resembles in body shape, by presence of blunt fingertips (fingertips triangular in shape in *U. taprobanicus*).

Colour Dorsum dark brown or black marbled with yellow or orange (Figs 1 & 2). Chin and chest area marbled with light brown. Throat of male black; throat of female mottled with brown. Belly white.

Habits Terrestrial, nocturnal and fossorial species that hides in loose, moist soil by day. Burrows efficiently, using shovel-shaped metatarsal tubercles in hindfeet as 'spades'. Usually seen in the rainy monsoon seasons. Invades heaps of sand kept in compounds brought for construction purposes. Breeding takes place during the monsoon rains, when males call from banks of streams, ponds, drains and paddy fields. When disturbed, adults inflate bodies like balloons.

Fig. 1 Dorsolateral aspect

Habitat and Distribution Uncommon frog usually found in dry forests, shrubland, plains, home gardens and agricultural areas. Widely distributed in lowland intermediate and dry zones.

Status Native. Extralimital: India, Nepal and Pakistan.

IUCN Red List Category Least Concern.

Fig. 2 Lateral aspect

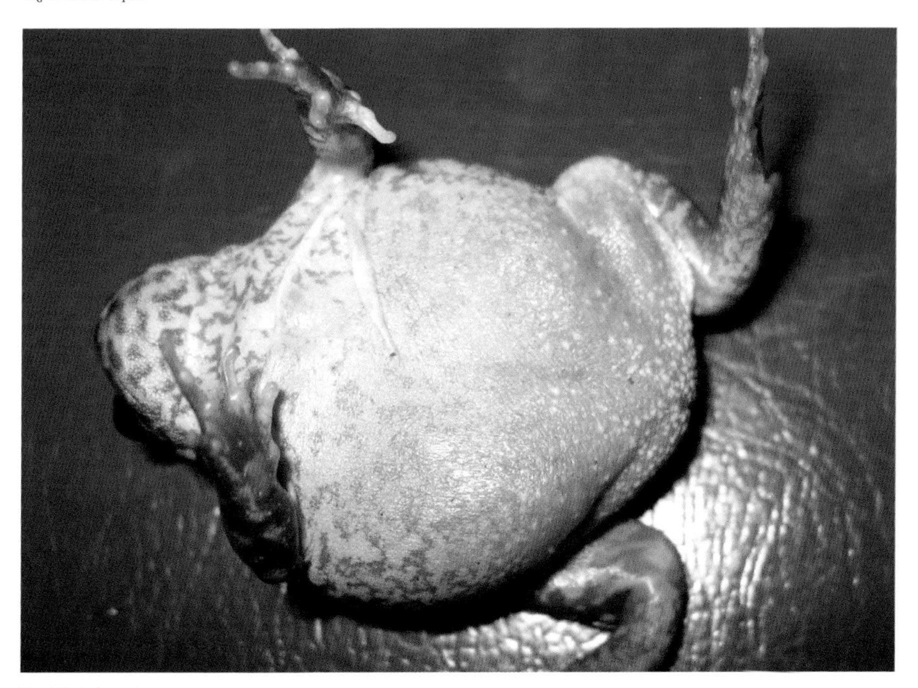

Fig. 3 Ventral aspect

Common Bull Frog *Uperodon taprobanicus*

(Visituru ratu mädiyã)

First described as *Kaloula pulchra taprobanica* by H.W. Parker in 1934. The genus *Kaloula* was later included in the genus *Uperodon*, and thus the current name is *Uperodon taprobanicus*.

Size SVL 30–65mm.

Identification Features Medium-sized frog with globose, teardrop-shaped body and triangular-shaped fingertips. Skin moist with numerous scattered tubercles on body. Head broader than long. Canthal ridge indistinct and loreal region concave. Tympanum not visible and supratympanic fold present. (Figs 1 & 2). Belly smooth or granular, and throat strongly granular. Toes one-third webbed and obtusely pointed. Toes and fingers depressed and edges have dermal fringes. Differs from *U. systoma*, the species it resembles most in body shape and by presence of triangular-shaped fingertips (blunt fingertips in *U. taprobanicus*).

Colour Although the species has many colour variations, an asymmetrical brick-red pattern on a greyish-black body is the most common colour pattern (Figs 1 & 2) from posterior edge of eye to base of upper arm. Belly spotted or marbled with grey or light brown mottling (Fig. 3).

Habits Can 'climb' rough walls and trees and occasionally may come inside houses. Characteristic call resembles bellowing of a bull. When cornered, inflates body and sinks head into lower jaw.

Habitat and Distribution Common burrowing species that hides by day under decaying logs, trees, rock crevices and leaf litter, or inside humus or soil. Commonly found in human habitats, agricultural fields, plantations and forest borders. Widely distributed all over the island from sea level to about 650m above.

Status Native. Extralimital: India and Nepal.

IUCN Red List Category Least Concern.

Fig. 1 Dorsolateral aspect

Fig. 2 Lateral aspect

Fig. 3 Ventral aspect

Family Nyctibatrachidae (Wrinkled Frogs)
This family of approximately 40 species, grouped in three subfamilies (Astrobatrachinae, Lankanectinae and Nyctibatrachinae), is restricted to South India and Sri Lanka. The subfamilies Astrobatrachinae and Nyctibatrachinae are restricted to the Western Ghats of South India, while the subfamily Lankanectinae is restricted to Sri Lanka. The Lankanectinae are characterized by heavily wrinkled bodies (transverse folds) and by two fang-like structures (odontoid processes) on the lower jaw. Adult frogs also possess a lateral line system, making them well adapted to aquatic life. They have indistinct tympanums and fully webbed hindlimbs, but lack any digital discs. The Lankanectinae comprise only two species, in the genus *Lankanectes*.

Genus *Lankanectes*
Because the subfamily Lankanectinae consists only of the genus *Lankanectes*, the characteristics of the subfamily apply to the genus as well. *Lankanectes* is the only group of frogs in Sri Lanka to have tooth-like processeses on the lower jaw, and this feature has also earned them the name 'Sri Lanka fanged frogs'. The adult frogs can reach a SVL of 70mm. Both species are completely aquatic, occurring in still and slow-flowing waters in the wet and intermediate zones of Sri Lanka.

Corrugated Water Frog *Lankanectes corrugatus*

(Vakaräli diya mädiyã)

First described as *Rana corrugata* by W. C. H. Peters in 1863. Transferred from *Rana* to *Limnonectes* by Alan Dubois in 1987, and to *Lankanectes* by Dubois and Ohler in 2001.

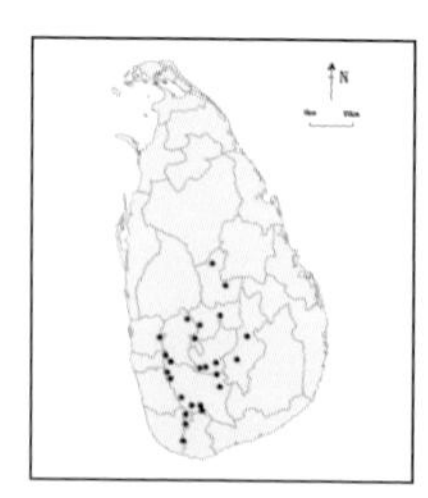

Size SVL 32–72mm.

Identification Features Small to medium-sized frog with transverse folds on body (wrinkled skin), moist tuberculated skin and fully webbed feet. Head longer than broad. Snout rounded when viewed dorsally. Canthal ridge indistinct and loreal region oblique. Tympanum not distinct and supratympanic fold not prominent. Small, white-tipped tubercles on upper eyelids and lateral sides of body, around vent and on both limbs. Fingers not webbed or rudimentarily webbed. Tips of fingers and toes rounded. Outer edge of fifth toe and inner edge of first toe have cutaneous fringe. Differs from *L. pera* by dark spots on throat and ventral sides of limbs (*L. pera* lacks spots on throat and ventral sides of limbs), and smooth throat (*L. pera* has granular throat).

Colour Dorsum brownish or brownish-orangish-brown with black blotches and spots, occasionally with mid-vertebral yellow or cream line (Figs 1 & 2). Some individuals have dark brown or brick-red, circular patches on dorsal side of body. Interorbital area has yellow transverse band. Limbs with or without black cross-bars. Prominent dark brown or black stripe between eye and forelimb. Belly light pinkish-white, and throat and limbs in some individuals mottled with brown (Fig. 3).

Habits Essentially an aquatic frog that occasionally comes to land close to water bodies. Characteristic call that is similar to a bird's whistle. Known to call by both day and night. Hides in loose mud when disturbed.

Habitat and Distribution Inhabits streams, marshes, pools in forests, areas near human habitation and paddy fields. Common in wet zone, including western, southern and central Sri Lanka, up to 1,525m above sea level.

Status Endemic.

IUCN Red List Category Least Concern.

Fig. 1 Dorsolateral aspect, with circular patches on dorsal side of body indicated by top arrow, and rounded finger- and toe-tips by bottom arrow

Fig. 2 Dorsolateral aspect

Fig. 3 Ventral aspect

KNUCKLES CORRUGATED WATER FROG *Lankanectus pera*

(Dumbara vakarali mädiyã)

First described as *Lankanectus pera* by G. Senevirathne, V. A. M. P. K. Samarawickrama, N. Wijayathilaka, K. Manamendra-Arachchi, G. Bowatte, D. R. N. S. Samarawickrama and M. Meegaskumbura in 2018.

Size SVL 35–66mm.

Identification Features Small to medium-sized frog with transverse folds (wrinkles) on body, moist tuberculated skin and webbed feet. Head dorsally flat and body stout. Snout rounded when viewed laterally and dorsally. Canthal edges indistinct and loreal region convex. Tympanum not visible. Dorsal surface of head and body covered with corrugations. Tips of digits rounded. Webbing on fingers absent, and toes fully webbed. Corrugations and glandular warts on dorsal surfaces of legs. (Fig. 1). Chest, belly and ventral parts of thighs smooth. Differs from *L. corrugatus* (bears spots on throat and ventral sides of limbs) by spotless throat and ventral sides of limbs, and granular throat (*L. corrugatus* has smooth throat) (Fig. 2).

Colour Dorsal surface light to chocolate-brown with unequal dark patches and small pale spots. Light brown bar with dark edges on interorbital area. Glandular warts white tipped. White tubercles on throat. Edge of upper lip, toes, feet and flanks uniformly grey. Chest, belly and ventral side white.

Habits Nocturnal and aquatic species. Adults found under rocks and in rock crevices. Tadpoles found in deeper regions of streams where decaying vegetation is gathered.

Habitat and Distribution Restricted to clear, shallow and slow-flowing, pristine streams with sand and rock-strewn substrates in closed canopy submontane and montane forests (elevation >1,000m) of the Knuckles Mountain Range, including Dothalugala, Knuckles Peak and Riverston.

Status Endemic.

IUCN Red List Category Critically Endangered.

Fig. 1 *Dorsolateral aspect*

Fig. 2 *Ventral aspect*

Family Ranidae (True Frogs)
Ranids, or 'true frogs', are characterized by their smooth, moist skin, distinct tympanum, powerful hindlimbs and webbed toes. Most have horizontal pupils, and they feature indirect development with a free-living tadpole that has keratinous mouthparts. There are 424 ranid species in 25 genera distributed all over the world, except southern South America, Antarctica and Madagascar. Sri Lanka is home to three species in two genera (*Hydrophylax* and *Indosylvirana*), of which all are endemic to the island and are closely associated with water. All Sri Lankan ranids are characterized by a dorsolateral ridge that runs along the body from the snout to the vent, a rictal gland at the back of the mouth, a distinct tympanum and round digital discs.

Genus *Hydrophylax*
Hydrophylax comprises four species of small to medium-sized frogs that are distributed in South and Southeast Asia. They are characterized by a robust body, granular skin in the posterior part of the abdomen and thick, well-developed dorsolateral fold and lateral glandular fold that extends from the back of the mouth to the groin. A single species of *Hydrophylax* has been recorded in Sri Lanka.

Sri Lanka Wood Frog *Hydrophylax gracilis*

(Lankã bädi mädiyã)

First described as *Rana gracilis* by J. L. C. Gravenhorst in 1829. After several changes to the species name (*Lymnodytes macularius, Hylorana macularia* and *Rana macularia*) and the generic name (*Rana gracilis* and *Hylarana gracilis*), the species is currently known as *Hydrophylax gracilis.*

Size SVL 36–55mm.

Identification Features Robust, small to medium-sized frog. Head broader than wide. Snout subovoid in dorsal view and rounded in lateral view. Loreal region acute and concave with rounded canthal ridge. Skin of snout, between eyes, sides of head and anterior part of dorsum shagreened; posterior part of back, and upper and lower parts of flanks shagreened and sparsely granular. Dorsal parts of forelimbs smooth. Thighs, tibias and tarsi without glandular warts and horny spinules. Ventral part of throat and anterior part of belly smooth; posterior parts of belly and thighs shagreened and sparsely granular. Can be distinguished from *Indosylvirana* species by granular skin in posterior part of abdomen (smooth skin in *Indosylvirana*), thick, well-developed dorsolateral fold from eye to vent (weakly developed in *Indosylvirana*) (Fig. 1a), distinct whitish lateral glandular fold that extends from back of mouth to groin (absent in *Indosylvirana*) (Fig. 1b), longitudinal stripes on shank (stripes on shank run across in *Indosylvirana*) (Fig. 2), and weakly developed, rounded digital discs (well-developed, rounded digital discs in *Indosylvirana*).

Colour Dorsum uniformly light brown to orangish-brown with two black stripes. Tympanum light brown. Upper lip has white stripe, continuing through rictal gland to above arm insertion. Thighs have dark longitudinal lines and backs of thighs marbled with black patches. Flanks dark brown and groin yellowish-grey. Throat and limbs light grey. Feet and webbing dark grey. (Figs 1 & 2). Belly white or light yellow.

Habits Lives close to water sources, on rocks and low vegetation, including in forests, paddy fields and anthropogenic habitats. Usually hides well in places that are very difficult to locate. Mainly calls at 7.30

to 10.30 p.m., and sometimes also during the day.

Habitat and Distribution Primary habitat is open wetland sites in both urban areas and secondary forests. Widely distributed in Sri Lanka and common in wet, intermediate and dry zones at 28–1,250m above sea level.

Status Endemic.

IUCN Red List Category Least Concern.

Fig. 1 Dorsalateral aspect, with fold from eye to vent indicated by top arrow, glandular fold from back of mouth to groin by bottom arrow

Fig. 2 Dorsolateral aspect, with longitudinal stripes on shank indicated by arrow

Genus *Indosylvirana*
Indosylvirana comprises 12 species of small to medium-sized frogs that are distributed in South and Southeast Asia. They are characterized by granular or wrinkled skin in the posterior part of the abdomen, and thin and well-defined dorsolateral folds with a pale or similar colour. Two species are known to occur in Sri Lanka, both of which are endemic.

Sri Lanka Golden-backed Frog *Indosylvirana serendipi*

(Ranwan bandi mädiya)

First described as *Hylarana serendipi* by S. D. Biju, S. Garg, S. Mahony, N. Wijayathilaka, G. Senevirathne and M. Meegaskumbura in 2014. Some members of the genus *Hylarana* were later placed in the genus *Indosylvirana*, and thus the current name is *Indosylvirana serendipi*.

Size SVL 30–45mm.

Identification Features Small, robust frog. Head small and longer than wide. Snout sub-elliptical in dorsal view and rounded in lateral view. Loreal region concave with rounded canthal ridge. Both fingers and toe-tips dorsoventrally compressed. Skin of snout, between eyes, sides of head, anterior part of dorsum, posterior part of back and upper parts of flanks granular. Upper eyelids in particular are more prominently granular. Lower parts of flanks smooth. Dorsal regions of forelimbs smooth. Thighs, tibias and tarsi have glandular warts in longitudinal lines bearing horny spinules. Ventral side of throat smooth and belly shagreened. Can be distinguished from H. *gracilis* by smooth skin in posterior part of abdomen (granular skin in H. *gracilis*), thin and weakly developed dorsolateral fold from eye to vent (thick and well developed in H. *gracilis*), stripes on shanks that run across (stripes runs longitudinally in H. *gracilis*), well-developed, rounded digital discs (weakly developed, rounded digital discs in H. *gracilis*) and absence of distinct whitish lateral glandular fold that extends from back of mouth to groin (present in H. *gracilis*) (Figs 1 & 2). Differs from I. *temporalis* by smaller size, slender habitus and absence of supratympanic fold (present in I. *temporalis*).

Colour Dorsum uniform light to dark brown with small black specks. Tympanum and surrounding area dark brown. Upper lip has gold to metallic-yellow stripe continuing through yellowish-white rictal gland to above arm insertion. Iris reddish-brown with golden specks and dark patches on either side. Flanks light yellowish-grey. Limbs dorsally light brown with light grey cross-bands. Throat and limbs light grey. Feet and webbing dark grey. (Figs 1 & 2). Ventral side white with brown speckles and orangish thighs.

Habits Nocturnal species.

Habitat and Distribution Occurs on banks of streams and in marshy areas, lowland rainforests and plantations. Distributed in lowland wet zone of Sri Lanka up to 1,000m above sea level. Common in Sinharaja (Kudawa, Lankagama, Enasalwatta, Runakanda and Morningside) and Kithulgala.

Status Endemic.

IUCN Red List Category Near Threatened.

Fig. 1 Dorsolateral aspect, with well-developed, rounded digital discs indicated by arrow

Fig. 2 Dorsolateral aspect

Günther's Golden-Backed Frog *Indosylvirana temporalis*

(Sulaba bandi mädiyā)

First described as *Hylorana temporalis* by A. C. L. G. Günther in 1864. After a number of changes to the genus (that is, through *Rana temporalis*, *Hylarana temporalis* and *Sylvirana temporalis*), the current name is *Indosylvirana temporalis*.

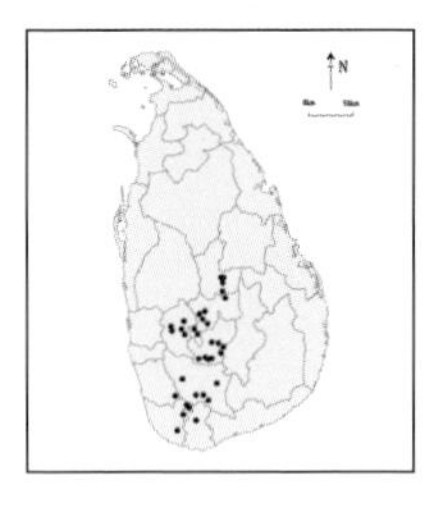

Size SVL 38–80mm.

Identification Features Medium to large, robust frog. Head longer than broad. Snout pointed when viewed laterally. Nostrils nearer to tip of snout than to eye and located laterally. Canthal ridge angular and loreal region concave. Tympanum distinct, rounded and vertically oval. Toes fully webbed. Dorsum finely granular. Tubercles on vent, thighs, tibias and heels larger and more elevated. Ventral sides of thighs have smooth granules. Can be distinguished from *H. gracilis* by smooth skin in posterior part of abdomen (granular skin in *H. gracilis*), thin and weakly developed dorsolateral fold from eye to vent (thick and well developed in *H. gracilis*), stripes on shanks that run across (stripes run longitudinally in *H. gracilis*), well-developed, rounded digital discs (weakly developed, rounded digital discs in *H. gracilis*), and absence of distinct whitish lateral glandular fold that extends from back of mouth to groin (present in *H. gracilis*). Differs from *I. serendipi* by larger size, robust build and presence of supratympanic fold (absent in *I. serendipi*).

Colour Dorsal side golden-brown, with or without some dark brown patches. Loreal and temporal regions brown to blackish-brown. Upper lip bright yellow or golden in colour. Limbs have distinct dark brown cross-bars (Figs 1 & 2). Belly light yellow or speckled with dark brown (Fig. 3).

Habits Stream-dwelling species that can be commonly seen on boulders adjacent to streams and pools in forests. Active by day and night, and can be heard vocalizing both day and night. Also capable of climbing trees to a certain extent.

Habitat and Distribution Common in forests, but can also be seen close to water sources in anthropogenic habitats such as tea and rubber estates. Widely distributed and common in wet zone at 60–1,830m above sea level.

Status Endemic.

IUCN Red List Category Near Threatened.

Fig. 1 Dorsolateral aspect, with supratympanic fold indicated by arrow b, distinct dark brown cross-bars by arrow a

Fig. 2 Dorsolateral aspect

Fig. 3 Ventral aspect

Family Rhacophoridae (Afro-Asian Tree-frogs)
Rhacophorids are small to large (maximum 120mm) frogs with enlarged toe-pads, flattened bodies, large eyes with horizontal pupils and relatively long limbs. They display both indirect and direct development. The free-living tadpoles have keratinous mouthparts. The family comprises the two subfamilies Buergeriinae and Rhacophorinae, with 431 species in 20 genera. Rhacophorids are distributed in Sub-Saharan Africa, Madagascar, and South, East and Southeast Asia. The majority are arboreal frogs, but some are also found on the ground. The Rhacophoridae is the most diverse family of frogs in Sri Lanka, with 83 species and 82 endemic species. In Sri Lanka, it is represented by the three genera *Polypedates*, *Pseudophilautus* and *Taruga*.

Genus *Polypedates*
Polypedates are medium-sized tree frogs measuring 45–65mm (SVL). They are mainly arboreal, and are distributed in South and Southeast Asia. They are characterized by their robust bodies, large eyes with horizontal pupils, elongated limbs with expanded fingertips, and free-living tadpoles with keratinous mouthparts. They place their eggs in foam nests that are attached to surfaces above water bodies. Once the eggs hatch the tadpoles fall into the water, where they live until metamorphosis. Three species are known from Sri Lanka, of which two are endemic.

Common Hour-glass Tree-frog *Polypedates cruciger*
(Sulaba pahimbu gas mädiyã)

First described as *Polypedates cruciger* by Edward Blyth in 1852. After several changes to the generic name (including to *Rhacophorus cruciger* and *R. c. cruciger*), it is currently known as *Polypedates cruciger*.

Size SVL 50–90mm.

Identification Features Medium to large frog with elongated body, long hindlegs and forelimbs, medially webbed toes, digits bearing distinct circular discs, bluntly pointed snout, round or oval tympanum, distinct supratympanic fold and concave loreal region. Both edges of digits have cutaneous fringes. Skin on head strongly attached to skull (co-ossified). Dorsum, chin and chest smooth. Belly and undersides of thighs granular. Resembles *P. maculatus*, but can be distinguished by hourglass mark on dorsum and co-ossified skin on head (vs absence of hourglass mark on dorsum and co-ossified skin on head in *P. maculatus*).

Colour Dorsum greenish-brown, dark brown or light yellow, with characteristic hourglass-shaped marking that extends from mid-level of eyes to mid-dorsum (Figs 1 & 2). In some individuals dorsal side is light brown with symmetrically arranged dark brown markings. Loreal and temporal regions brown. Dorsum of limbs has black or brown cross-bars. Backs of thighs uniformly brown with dark brown spots. Venter white.

Habits Nocturnal and arboreal species, common around human habitations. Lays eggs on branches or any surface above a water body. Tadpoles develop in water.

Habitat and Distribution Known from many locations in wet, intermediate and dry zones up to 1,500m above sea level. Common around human habitation in wet zone forests. By day, hides under loose bark, and among sheaths and leaves in banana groves and other vegetation.

Status Endemic.

IUCN Red List Category Least Concern.

Fig. 1 Dorsolateral aspect

Fig. 2 Dorsolateral aspect, with hourglass-shaped mark indicated by arrow

Spotted Tree-frog *Polypedates maculatus*

(Pulli gas mädiyã)

First described as *Hyla maculata* by John Edward Gray in 1830. After several changes to the generic name (including to *Bürgeria maculata, Hyla reynoudi, Polypedates maculatus, Rhacophorus maculatus, R. maculatus, R. acanthostomus* and *R. leucomystax maculatus*), it is currently known as *Polypedates maculatus*.

Size SVL 34–70mm.

Identification Features Medium-sized frog with elongated body, long hindlegs and forelimbs, medially webbed toes, digits bearing distinct circular discs, bluntly pointed snout, round or horizontally oval tympanum, distinct supratympanic fold and concave loreal region. Dorsum smooth, and chin and chest smoothly granular. Belly and undersides of thighs granular. Resembles *P. cruciger*, but can be distinguished by absence of hourglass mark on dorsum and co-ossified skin on head (vs presence of hourglass mark on dorsum and co-ossified skin on head in *P. cruciger*) (Figs 1 & 2).

Colour Dorsal colour highly variable, from olivaceous to chestnut, sometimes brownish-yellow or grey, with scattered dark spots (Figs 1 & 2). Loreal region black or dark brown. Dorsal sides of limbs have brown cross-bars. Belly white (Fig. 3).

Habits Several individuals may be found in one place during the day. Male and female make a spawn nest and female lays eggs inside this. Squirts urine if caught and can leap to a distance of 2–3m as an escape strategy.

Habitat and Distribution Common species in human habitations and forests. In dry zone often encountered in houses and bathrooms. By day hides in banana groves and cool, moist places in anthropogenic habitats. Widely distributed in dry and wet zone lowlands up to 450m.

Status Native. Extralimital: India, Bhutan, Nepal and Bangladesh.

IUCN Red List Category Least Concern.

Fig. 1 Dorsolateral aspect

Fig. 2 Lateral aspect

Fig. 3 Ventral aspect

Ranwella's Spined Tree-frog *Polypedates ranwellai*

(Ranwellage anga gas mediya)

Described as *Polypedates ranwellai* by Mendis Wickramasinghe, Amith Munindradasa and Prithiviraj Fernando in 2012.

Size SVL 41–65mm.

Identification Features Medium to large frog with elongated body, truncated snout in dorsal aspect, rounded snout in lateral and ventral aspects, rounded canthal edges, concave loreal region, convex interorbital space, long hindlegs and forelimbs, medially webbed toes, digits bearing distinct circular discs, oval-shaped vertical tympanum, curved supratympanic fold from back of eye to end of jaw, four dorsal spines on back of head and protrusion at proximal end of jaw. Skin on head co-ossified with skull. Blunt calcar at tibiotarsal articulation. Can be easily distinguished from all other species of *Polypedates* in Sri Lanka by four dorsal spines on back of head and protrusion at proximal end of jaw (Fig. 1).

Colour Dorsal part of head and dorsum have very small black spots on bright luminous yellow to brown or ashy-white background. Lower parts of flanks marbled with black and off white. Forelimbs have faint black stripes on yellow and small black spots all over. Tibias and tarsals have three faint black stripes on yellow and scattered small black spots. Posterior parts of femurs marbled with black and whitish-yellow, and ventral part white with shaded black. (Figs 1 & 2).

Habits Nocturnal and arboreal. Perches on branches about 1.5m above the ground.

Habitat and Distribution Occurs in forests and forest edges and known from a few locations in lowland wet zone of Sri Lanka, including Gilimale, Runakanda and Dellawa.

Status Endemic.

IUCN Red List Category Endangered.

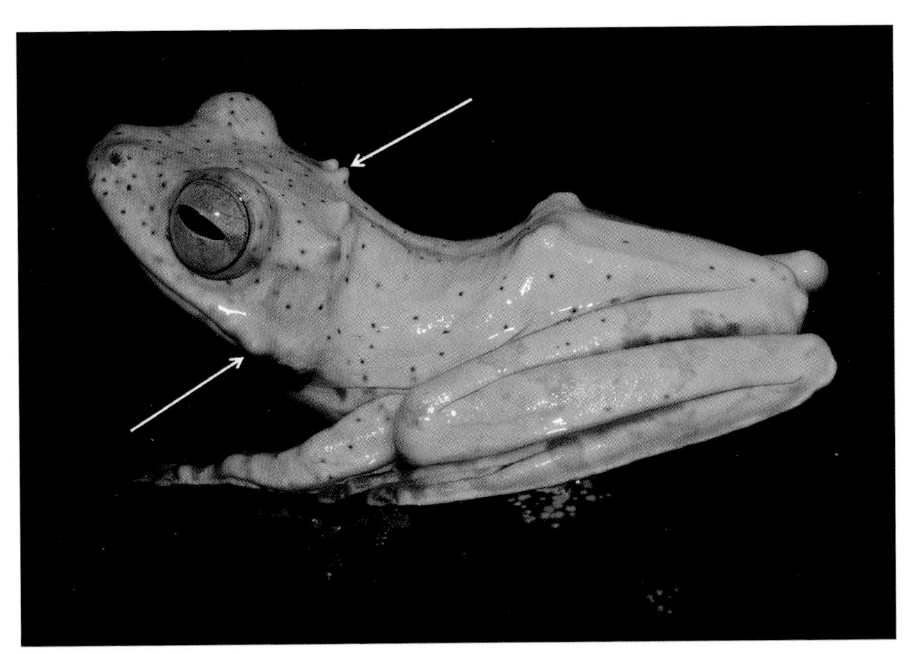

Fig. 1 Lateral aspect, with spines on back of head indicated by top arrow, protrusion at proximal end of jaw by bottom arrow

Fig. 2 Dorsolateral aspect

Genus *Pseudophilautus*
Pseudophilautus are small to medium-sized (20–50mm SVL) frogs that are mostly seen in wet leaf litter, shrubs and trees. They are commonly known as 'shrub frogs' or 'Oriental shrub frogs'. This is the most diverse genus of frogs in Sri Lanka, with 77 known species, all of which are endemic to Sri Lanka. Seventeen are thought to already be extinct on the island. *Pseudophilautus* species do not have a tadpole stage but instead display the phenomenon of 'direct development' – larval development that occurs within the eggs so that hatchlings resemble miniature adults.

Labugama Shrub Frog *Pseudophilautus abundus*

(Labugama pañduru mädiya)

First described as *Philautus abundus* by Kelum Manamendra-Arachchi and Rohan Pethiyagoda in 2005. Sri Lankan and Indian *Philautus* species were later placed in the genus *Pseudophilautus*, and thus its current name is *Pseudophilautus abundus*.

Size SVL 25–37mm.

Identification Features Small frog with slender body, dorsally concave head, pointed or oval snout in lateral aspect, rounded canthal edges, flat interorbital space, distinct tympanum and supratympanic fold, lateral dermal fringe on fingers and fully webbed toes. Both loreal region and internarial space concave. Tympanum oval, and vertically arranged. Lingual papilla present. Snout, interorbital area, and sides of head smooth. Dorsum has horn-like spinules. Upper flanks smooth and lower flanks granular. Dorsal parts of forelimbs, thighs, shanks and feet smooth. Fine granules on throat and chest. Belly and

Fig. 1 Dorsalateral aspect, with infraorbital patch indicated by arrow

undersides of thighs granular. Resembles *P. procax*, but differs from it by fully webbed toes in *P. abundus* (vs medially webbed toes in *P. procax*). (Figs 1 & 2).

Colour Dorsum pale brown with dark brown markings. Yellowish or pale mid-dorsal stripe may be present from tip of snout to vent in some individuals. Yellow infraorbital patch. Chin and chest pale brownish-yellow. Colours of dorsum and venter meet in well-defined zone on flanks. Limbs and thighs dorsally dark brown. (Figs 1 & 2). Abdomen pearly-white to pale yellow.

Habits Nocturnal species. Males can be seen vocalizing perched on leaves about 1m above the ground.

Habitat and Distribution Habitat generalist that can be seen in forest edges, and both open and closed canopy forests. Abundant near streams and marshy areas in forests and cardamom plantations. Widely distributed in lowland forests of south-west wet zone of Sri Lanka at 70–300m above sea level. Known from Kanneliya, Dediyagala, Labugama, Yagirala and Kithulgala.

Status Endemic.

IUCN Red List Category Not Evaluated.

Fig. 2 Dorsolateral aspect

THWAITES SHRUB FROG *Pseudophilautus adspersus*

(Thwatsige panduru mädiya)

First described as *Ixalus adspersus* by A. C. L. G. Günther in 1872. After changes to the generic name (that is, to *Rhacophorus adspersus*, Ahl, 1931, and *Philautus adspersus*, Inger, 1931), currently documented as *Pseudophilautus adspersus*. Known only from two specimens and last reported in around 1886. Now thought to be extinct, as recent extensive field surveys of the amphibian fauna of Sri Lanka have failed to rediscover it.

Status Endemic.

IUCN Red List Category Extinct.

HORTON PLAINS SHRUB FROG *Pseudophilautus alto*

(Mahaeli panduru mädiyã)

First described as *Philautus alto* by Kelum Manamendra-Arachchi and Rohan Pethiyagoda in 2005. Generic name later changed to *Pseudophilautus*, and thus the current name is *Pseudophilautus alto*.

Size SVL 17–28mm.

Identification Features Small frog with distinct tympanum and supratympanic fold, rounded canthal edges, obtusely pointed snout laterally, medially webbed feet, dorsum with horny spinules, calcar (spine) at tibiotarsal articulation (heel) and dark brown lateral stripe running from loreal region to base of upper arm. Body elongated and head dorsally convex. Loreal region concave. Internarial space concave. Tympanum oval, vertical. Skin on head not co-ossified with skull. Lateral dermal fringes on fingers. Toes medially webbed. Snout, interorbital area, dorsum and upper flanks have horn-like spinules in male. Dorsum warty in female. Sides of head smooth. Lower flanks granular. Dorsal parts of forelimbs, thighs, shanks and feet smooth. Throat, chest, belly and undersides of thighs granular. Resembles *P. cuspis*, *P. malcolmsmithi*, *P. rugatus* and *P. zorro*. Differs from *P. cuspis* and *P. zorro* by presence of rounded canthal edges and row of tubercles placed in ')(' shape on back (vs presence of sharp canthal edges and row of tubercles placed in ')(' shape on back in *P. cuspis* and *P. zorro*). (Figs 1 & 2).

Colour Colours and patterns highly variable. Generally dorsal colour varies from olive-brown to grey to dark brown, with cream or white vertebral line from tip of snout to vent (Figs 1 & 2). Dorsal and lateral parts of limbs ashy-light brown with dark olive-green cross-bars. Limbs have dark olive-brown cross-bars. Belly a mixture of brown, light yellow and rose colours, and chin area darker with white line medially from lower lip to mid-belly (Fig. 3).

Fig. 1 Dorsolateral aspect

Habits Can be seen on leaf litter on the forest floor, on moss and fern-covered rocks, among *Strobilanthus* roots, and at bases of other moss-covered roots of large trees. Ground-nesting, direct-developing species that is active by both day and night. During the night, adult males vocalize, perching on branches and leaves. During wet or humid weather, males call even in daytime within shrubs. Mature females can also be seen on shrubs.

Habitat and Distribution Habitat generalist that is found mostly on shrubs (usually less than 0.3–2m above the ground), in anthropogenic habitats and forest edges. Recorded from the Central Hills at 1,890–2,135m, including at Horton Plains, Ambewela, Pattipola, Hakgala, Sri Pada and Piduruthalagala.

Status Endemic.

IUCN Red List Category Endangered.

Fig. 2 Lateral aspect

Fig. 3 Ventral aspect

Asanka's Shrub Frog *Pseudophilautus asankai*

(Asankage panduru mädiyã)

First described as *Philautus asankai* by Kelum Manamendra-Arachchi and Rohan Pethiyagoda in 2005. Generic name later changed to *Pseudophilautus*.

Size SVL 18–34mm.

Identification Features Small frog with indistinct tympanum and feebly defined supratympanic fold, rounded canthal edges, obtusely pointed snout when viewed laterally, medially webbed feet, dorsum with horny spinules and granular belly. Loreal region concave. Interorbital and internarial spaces concave or flat. Body elongated and head dorsally flat or convex. Snout and interorbital area smooth. Sides of head smooth or with glandular warts. Dorsum and upper flanks shagreened in both sexes and with horn-like spinules in male. Throat and chest smooth or shagreened. Dorsal sides of forelimbs have glandular warts. Thighs, shanks and feet dorsally smooth. (Figs 1 & 2). Belly and undersides of thighs granular (Fig. 3). Resembles *P. pleurotaenia* and *P. auratus*. Differs from *P. auratus* by absence of dermal fringes on fingers, presence of horny spinules on dorsum, glandular and warty dorsum, and dorsally convex head (vs dermal fringes on fingers present, horny spinules on dorsum absent in male (dorsum shagreened) and head dorsally flat, in *P. auratus*). Differs from *P. pleurotaenia* by distinct supratympanic fold, no dermal fringe on fingers (vs indistinct supratympanic fold, and dermal fringe on fingers in *P. pleurotaenia*).

Colour Dorsal colour highly variable, from pale reddish-brown with ashy mottling, to pale green or light blue or yellow. Belly and ventral aspect of limbs bright yellow. Some individuals have spots on dorsum and limbs. Fingers, toes and feet yellow dorsally (Figs 1 & 2). Venter bright yellow (Fig. 3).

Habits Nocturnal and arboreal species that can be seen on the ground by day. Habitat generalist that could adjust to living in anthropogenic habitats. Usually, males perch on shrubs about 1.5–3m above the ground and call at night.

Fig. 1 Dorsolateral aspect

Habitat and Distribution Restricted to central Sri Lanka at 810–1830m, including at Hakgala, Piduruthalagala, Moray Estate, Dayagama and Agrapathana. Common in open canopy, including in low bushes, borders of tea plantations, vegetable plantations and anthropogenic habitats.

Status Endemic.

IUCN Red List Category Endangered.

Fig. 2 Lateral aspect

Fig. 3 Ventral aspect

Golden Shrub Frog *Pseudophilautus auratus*

(Ranwan panduru mediya)

First described as *Philautus auratus* by Kelum Manamendra-Arachchi and Rohan Pethiyagoda in 2005. Sri Lankan and Indian *Philautus* species were later placed in the genus *Pseudophilautus*, and thus the current name is *Pseudophilautus auratus*.

Size SVL 22–28mm.

Identification Features Small frog with distinct tympanum and supratympanic fold, rounded canthal edges, medially webbed feet, shagreened dorsum and lingual papilla. Body slender with dorsally flat head. Snout blunt in lateral aspect. Canthal edges rounded and internarial space concave. Loreal region and interorbital space flat. Snout, interorbital area, sides of head and dorsum shagreened. Upper flanks granular or smooth and lower flanks granular. Dorsal parts of forelimbs and feet shagreened or smooth. Thighs and shanks smooth. Throat and chest smooth or granular. Belly and ventral sides of thighs granular. Resembles *P. asankai* and *P. pleurotaenia*. Differs from *P. pleurotaenia* by distinct supratympanic fold, half-webbed toes, shagreened dorsum and head dorsally flat in lateral view (vs supratympanic fold indistinct, toes fully webbed, dorsum glandular warty or shagreened, and head dorsally convex in lateral view in *P. pleurotaenia*). Distinguished from *P. asankai* by dermal fringes on fingers and head dorsally flat (vs dermal fringes on fingers absent and head dorsally convex in *P. asankai*). (Figs 1 & 2).

Colour Highly variable, from pale yellow to orangish-brown. Dorsum and lateral sides of head pale yellow with dark brown spots. (Figs 1 & 2). Abdomen white or pale yellow with golden-yellow granules.

Habits Fairly uncommon habitat specialist with restricted distribution. Adult males usually seen perched about 1–1.5m above the ground on understorey shrubs, calling mainly at night.

Habitat and Distribution Only found in closed canopy rainforests, cloud forests, cardamom plantations within cloud forests, close to streams and marshy areas. Known from the Sinharaja Forest Reserve in Kudawa, Morningside, Handapan Ella Plains and Erathna at 510–1,270m.

Status Endemic.

IUCN Red List Category Endangered.

Fig. 1 Dorsolateral aspect

Fig. 2 Lateral aspect

Bambaradeniya's Shrub Frog *Pseudophilautus bambaradeniyai*

(Bambaradeniyage panduru mediya)

First described as *Pseudophilautus bambaradeniyai* by L. J. M. Wickramasinghe, D. R. Vidanapathirana, M. D. G. Rajeev, S. C. Ariyarathne, A. W. A. Chanaka, L. L. D. Priyantha, I. N. Bandara and N. Wickramasinghe in 2013.

Size SVL 17–20mm.

Identification Features Small, elongated frog with distinct tympanum and supratympanic fold, rounded canthal edges, medially webbed feet and tuberculated dorsum. Head large and dorsally convex. Snout truncated in lateral aspect and mucronate in dorsal aspect. Internasal space flat and interorbital space convex. Loreal region concave. Rudimentary webbing on all fingers. Subarticular tubercles prominent and oval in shape. Skin of snout (dorsally and laterally), lateral head, upper and lower parts of flanks, upper arms, forearms, hands and interorbital spaces smooth. Anterior dorsum has prominent tubercles and horny spinules, and posterior with prominent horny spinules. Supratympanic fold distinct. Legs smooth with few tubercles. Throat and chest weakly granular. Belly granular. Ventral sides of upper arms weakly granular to smooth, and forearms, thighs, legs and tarsi smooth. Resembles *P. rus*, from which it can be distinguished by truncated snout when viewed laterally and dermal fringes on posterior margins of lower arms (vs oval snout when viewed laterally and absence of dermal fringes on posterior margins of lower arms in *P. rus*). (Figs 1 & 2).

Colour Dorsally light to dark brown with blackish blotches. Large, dark brown marking covers much of anterior part of body. Blackish cross-band between eyes and prominent off-white vertebral stripe from tip of snout to vent that continues down hindlimbs symmetrically. Body colour lighter laterally and limbs dark brown. (Figs 1 & 2).

Habits Occupies the forest floor to shrubs about 1.5m above the ground.

Habitat and Distribution Found from lowland rainforests to lower montane rainforests at 750–1,400m. Known only from the Sripada (Peak Wilderness) Sanctuary.

Status Endemic.

IUCN Red List Category Critically Endangered.

Fig. 1 Dorsolateral aspect

Fig. 2 Lateral aspect

BLUE-THIGH SHRUB FROG *Pseudophilautus caeruleus*

(Nil kalawethi panduru mädiyã)

First described as *Philautus caeruleus* by Kelum Manamendra-Arachchi and Rohan Pethiyagoda in 2005. Sri Lankan and Indian *Philautus* species were later placed in the genus *Pseudophilautus*, and thus the current name is *Pseudophilautus caeruleus*.

Size SVL 17–20mm.

Identification Features Small frog with distinct tympanum and supratympanic fold, sharp canthal edges, medially webbed feet, depressed lingual papilla, and dorsum with glandular warts and horny spinules, and bluish colour in inguinal zone and thighs. Body slender, snout obtusely pointed in lateral aspect and loreal region concave. Interorbital space flat and internarial space flat or concave. Tympanum distinct and oval in shape. Snout, interorbital area, sides of head and upper flanks have glandular warts and horn-like spinules (females lack horn-like spinules on dorsum). Dorsal parts of forelimbs, thighs, shanks and feet smooth. Throat granular with scattered glandular warts. Chest, belly and undersides of thighs and lower flanks granular in male. Resembles *P. simba* and *P. semiruber*. Can be differentiated by dark brown lateral line from tip of snout to flanks, loreal region and tympanic region (below stripe) being light in colouration and blue/greyish pigments in inguinal zone (vs no such stripe, whole loreal region and tympanic region being dark and without blue or greyish pigments in inguinal zone in *P. simba* and *P. semiruber*). (Figs 1 & 2).

Colour Dorsum orangish-brown to dark browinish-grey-white with black spots and blotches. Unique bluish colour in inguinal zone and thighs. Dorsal parts of thighs and shanks light brown with four dark transverse bands on each. Throat, chest and webbing ventrally light brown. (Figs 1 & 2). Belly light brown with dark brown patches.

Habits Nocturnal species found among leaf litter and under logs by day. Males usually perch on branches of low shrubs 20–50cm above ground level.

Habitat and Distribution Can be seen in closed canopy habitats of both disturbed and undisturbed forests. Fairly uncommon species restricted to the Central Hills of Sri Lanka, including the Peak Wilderness Sanctuary, Moray Estate, Rajamale Estate, Bogawanthalawa and Rilagala at 810–1,650m.

Status Endemic.

IUCN Red List Category Endangered.

Fig. 1 Dorsolateral aspect

Fig. 2 Lateral aspect

HOLLOW-SNOUTED SHRUB FROG *Pseudophilautus cavirostris*

(Hirigadu panduru mädiyã)

First described as *Polypedates cavirostris* by A. C. L. G. Günther in 1869. After a number of changes in the generic name (that is, to *Rhacophorus cavirostris, Philautus cavirostris* and *Kirtixalus cavirostris*), the currently accepted name is *Pseudophilautus cavirostris*.

Size SVL 38–49mm.

Identification Features Small to medium-sized frog with distinct tympanum and supratympanic fold, sharp canthal edges, fully webbed feet, heavily tuberculated dorsum, tuberculated fringes on lower arms and feet (tarsal fold), calcar (spur) at tibiotarsal articulation (heel), and prominent conical tubercles around vent. Body stout and dorsal surface of head concave. Loreal region, and interorbital and internarial space, concave. Tympanum oval shaped. Dermal fringes on fingers. Dorsum, sides of head, snout and interorbital area have glandular warts. Upper flanks smooth and lower flanks granular. Dorsum of forelimbs, thighs, shanks and feet granular. Throat and chest smooth. Belly and ventral sides of thighs granular. Resembles *P. schmarda*, *P. decoris* and *P. mittermeiri*. Distinguished from *P. schmarda* by larger size, oval-shaped snout in lateral view and fully webbed toes (vs smaller size of SVL 17–30.0mm, obtusely pointed snout in lateral view and partially webbed toes in *P. schmarda*). Differs from *P. decoris* and *P. mittermeiri* in having fully webbed toes, and a concave snout in lateral aspect (vs half-webbed toes and flat snout when viewed in lateral aspect in *P. decoris* and *P. mittermeiri*).

Colour Dorsal part of head and body a mixture of dull grey, brown or olive-green with lighter patches. Dorsal sides of both upper and lower arms grey, or brownish with darker bands on dorsal sides. Two yellowish-green patches on interorbital

Fig. 1 Dorsolateral aspect

area and several light brown patches on mid-dorsum. Foot dorsally dark ash and posterior edge ashy-brown with dark brown patches. (Figs 1 & 2). Throat light yellowish with brown patches and belly light yellowish with reddish tinge.

Habits Nocturnal species in which adults mostly perch on shrubs about 0.3–2.0m above the ground. Frequents branches, mossy logs and tree trunks, mossy rock surfaces, tree-holes and crevices close to streams. Has the ability to merge well with its substrates.

Habitat and Distribution Uncommon habitat specialist found only in closed canopy rainforests of Sri Lanka. However, may occasionally be found even in heavily shaded humid home gardens. Widely distributed in wet zone forests at elevations of 200–1,000m. Known from many localities in central and southern Sri Lanka, including Hiyare, Sinharaja World Heritage Site (Kudawa, Lankagama), Polgolla, Gannoruwa, Ambagamuwa, Gampola, Pilimathalawa, Ratnapura, Akuressa, Haycock, Weddagala, Kitulgala, Kosmulla, Pathanagala, Kadugannawa and Pussellawa.

Status Endemic.

IUCN Red List Category Vulnerable.

Fig. 2 Lateral aspect

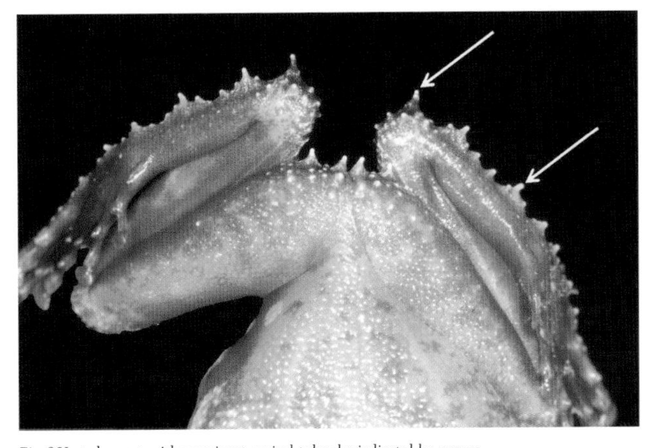

Fig. 3 Ventral aspect, with prominent conical tubercles indicated by arrows

Connif's Shrub Frog *Pseudophilautus conniffae*

(Connifge panduru mädiyã)

First described as *Pseudophilautus conniffae* by S. Batuwita, M. De Silva and S. Udugampala in 2019.

Size Size SVL 20–24mm.

Identification Features Small frog with distinct tympanum and supratympanic fold, rounded canthal edges, glandular dorsum with prominent warts and medially webbed feet. Snout rounded in lateral aspect and obtuse in dorsal aspect. Canthal ridges rounded. Loreal region concave. Interorbital region flat. Internarial region slightly convex. Tympanum oblique, and oval in shape. Snout, interorbital region, sides of head, and dorsum have isolated glandular warts. Dorsal areas of thigh, shanks and pes have a few scattered granules. Chest, belly and undersides of thighs heavily granular. Flanks have isolated, indistinct granules. Dorsal and ventral surfaces of forelimbs smooth. (Figs 1–3). Resembles *P. rus* and *P. silvaticus*. Differs from *P. silvaticus* by relatively acute snout, skin on head co-ossified with cranium, and no 'V'-shaped tubercular pattern on dorsum (vs skin on head not co-ossified with cranium, and 'V'-shaped tubercular pattern on dorsum in *P. silvaticus*). Differs from *P. rus* by having fourth toe webbing to penultimate subaricular tubercle on inner and outer sides (vs fourth toe webbing in between penultimate and antepenultimate subaricular tubercles on inner and outer sides), and black patches on posterior flanks, and anterior and posterior edges of thighs (Fig. 3) (vs black patches on anterior surfaces of thighs).

Colour Dorsum light brown with dark brown patches (Figs 1 & 2). Tympanum and loreal region dark brown, and tympanum bordered with lighter margin. Large, dark brown, square-shaped patch in anterior mid-dorsal region of body. Upper flank light brown to chestnut. Lower flank

Fig. 1 Dorsal aspect

spotted in female. Both dorsal and lateral regions of forelimbs and dorsal region of thighs, shanks and pes brown with dark brown cross-bars (Fig. 3).

Habits Habitat specialist that can be seen often in bamboo vegetation. Usually perches on leaves and bamboo branches less than 1m above the ground, and calls.

Habitat and Distribution Confined to rainforest areas in Galle and Matara districts, including Dediyagala, Kottawa-Kombala and Hiyare Forest Reserves at 80–300m.

Status Endemic.

IUCN Red List Category Endangered.

Fig. 2 Lateral aspect

Fig. 3 Femur with black spots and thighs with dark brown cross-bars, indicated by arrow

Sharp-snouted Shrub Frog *Pseudophilautus cuspis*

(Thiuynu hombu panduru mädiyã)

First described as *Philautus cuspis* by K. Manamendra-Arachchi and R. Pethiyagoda in 2005. The genus *Philautus* was later changed to *Pseudophilautus* (Laurent, 1943), and the current name is *Pseudophilautus cuspis*.

Size SVL 18–30mm.

Identification Features Small frog with distinct tympanum and supratympanic fold, sharply pointed snout when viewed laterally, sharp canthal edges, slender body, medially webbed feet and calcar at tibiotarsal articulation, granular belly and dark brown lateral stripe from tip of snout to base of upper arm. Head dorsally convex. Tympanum vertical and oval in shape. Sides of head smooth. Upper flanks smooth or with glandular warts, and lower flanks smooth or granular. Narrow dermal fringe on mid-dorsum from tip of snout to vent. Dorsal parts of forelimbs, thighs and feet smooth or with glandular warts. Shank with glandular folds and scattered with glandular warts. Throat smooth. Chest smooth or granular. Undersides of thighs granular. (Figs 1–3). Resembles *P. zorro* and *P. alto*. Distinguished from *P. zorro* by pointed elongated snout, dorsally convex head, flat loreal and internarial region, and lack of supernumerary tubercles on feet (vs pointed snout, supernumerary tubercles on feet, dorsally concave head, and concave loreal and internarial region in *P. zorro*). Differs from *P. alto* by sharp canthal edges, ')('-shaped rows of horn-like spinules on dorsum, flat loreal region and internarial region, and no dermal fringes on fingers and supernumerary tubercles on feet (vs rounded canthal edges, dermal fringes on fingers, supernumerary tubercles on feet, concave loreal region and internarial region, and no ')('-shaped rows of horn-like spinules on dorsum in *P. alto*).

Colour Dorsal colouration highly variable, from beige to dark brown. Dark brown stripe from snout to bases of upper limbs. Loreal and temporal regions black. Dorsal

Fig. 1 Dorsal aspect

sides of limbs have brown cross-bars, and thighs dark brown. (Figs 1–3).

Habits Active by both day and night. At night, males perch and vocalize from branches and leaves about ~0.3m above the forest floor. They merge with their substrates, making them difficult to see. Occurs on leaf litter, low vegetation, moss-covered rocks, tree trunks and the ground.

Habitat and Distribution Habitat specialist found only in closed canopy rainforests. Restricted to low-country wet zone, including Sinharaja, Morawaka, Deniyaya and Enasalwatta at 150–1,100m.

Status Endemic.

IUCN Red List Category Endangered.

Fig. 2 Dorsolateral aspect

Fig. 3 Lateral aspect, with sharply pointed snout indicated by arrow

Dayawansa's Shrub Frog *Pseudophilautus dayawansai*

(Dayawansage panduru mädiyã)

First described as *Pseudophilautus dayawansai* by L. J. M. Wickramasinghe, D. R. Vidanapathirana, M. D. G. Rajeev, S. C. Ariyarathne, A. W. A. Chanaka, L. L. D. Priyantha, I. N. Bandara and N. Wickramasinghe in 2013.

Size SVL 24–27mm.

Identification Features Small, elongated frog with distinct tympanum and supratympanic fold, rounded canthal edges, medially webbed feet, lightly tuberculated dorsum, blunt calcar (spur) at tibiotarsal articulation (heel), and dermal fringes on fingers. Head dorsally convex. Snout truncated laterally, mucronate dorsally and sub-elliptical ventrally. Internasal space and interorbital space concave. Loreal region concave. Tympanum vertical and oval in shape. Dermal fringes on insides of all fingers and both sides of all toes. Skin between eyes smooth with two symmetrically positioned ridges that do not connect mid-dorsally. Anterior dorsum smooth and posterior dorsum weakly tubercular (Fig. 1). Both upper and lower parts of flanks granular. Upper arms shagreened and weakly tubercular. Inner, outer and dorsal thighs smooth. Feet smooth. Legs and tarsi weakly tuberculated (Fig. 2). Resembles *P. sarasinorum*, from which it can be differentiated by single medially placed prominent tubercle on snout, smooth interorbital region and shagreened throat (vs smooth snout and interorbital region and throat with glandular warts in *P. sarasinorum*).

Colour Dorsum brown with dark brown blotches, and mostly covered with dark brown markings forming clear, semi-circular, blackish margin at anterior side. Its centre has a large, light brown triangle that is symmetrically placed on dorsum with base on sacral hump pointing towards head. Three prominent blackish-brown circular spots, and two between eyes. (Figs 1 & 2).

Habits Known to perch on high shrubs about 2m above ground level.

Habitat and Distribution Known only from montane forests at 1,550–1,900m above sea level in Sripada (Peak Wilderness) sanctuary of Sri Lanka.

Status Endemic.

IUCN Red List Category Critically Endangered.

Fig. 1 Dorsolateral aspect

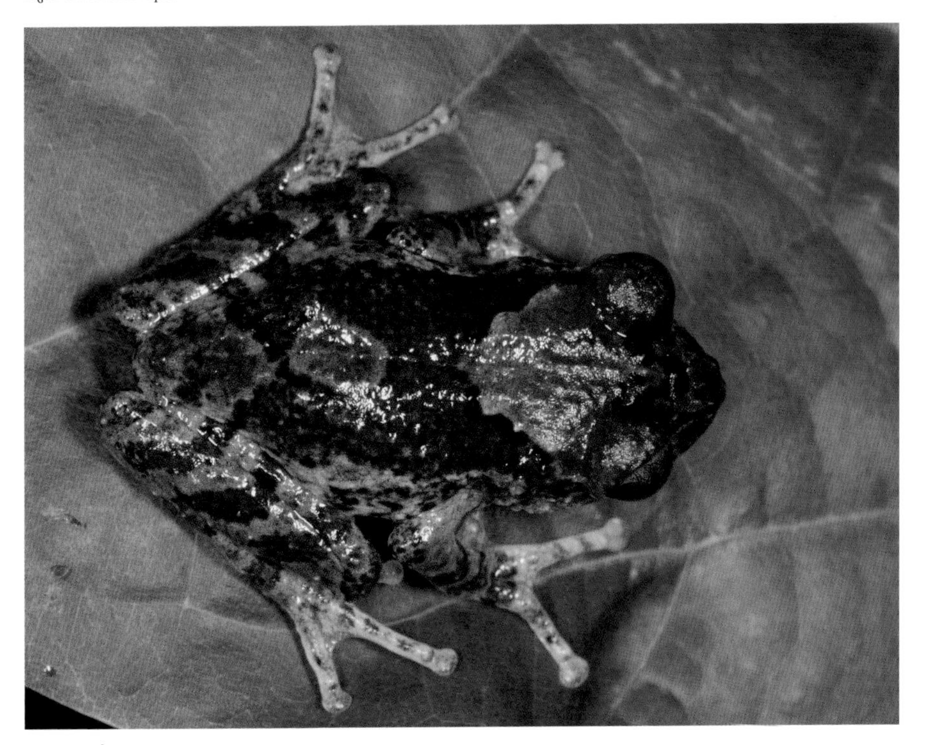

Fig. 2 Dorsal aspect

Elegant Shrub Frog *Pseudophilautus decoris*

(Bushana panduru mädiyã)

First described as *Philautus decoris* by Kelum Manamendra-Arachchi and Rohan Pethiyagoda in 2005. Sri Lankan and Indian *Philautus* species were later placed in the genus *Pseudophilautus*, and thus the current name is *Pseudophilautus decoris*.

Size SVL 18–24mm.

Identification Features Small frog with distinct tympanum and supratympanic fold, dermal fringes on margins of lower arms and tarsi (tarsal fold), rounded or sharp canthal edges, medially webbed feet, heavily tuberculated dorsum, prominent calcar (spur) at tibiotarsal articulation (heel), rounded lingual papilla and dermal fringes on fingers. Body stout with dorsally flat head. Loreal region concave. Interorbital space concave and internarial space concave or flat. Snout, interorbital area, sides of head, dorsum, flanks, and dorsal parts of forelimbs, thighs, shanks and feet have glandular warts. Throat glandular, warty or roughly granular. Ventral sides of thighs granular and not rough. Chest and belly granular and rough. Resembles *P. schmarda* and *P. mittermeieri*. Differentiated from *P. schmarda* by sheath-like, wavy fold on posterior margins of lower arms and feet, and cross-bars on limbs (vs row of distinct tubercles on posterior margins of lower arms and feet, and blotches on limbs in *P. schmarda*). Differs from *P. mittermeieri* by sharp, pointed snout in lateral aspect, smooth dorsal surface on forelimbs, thighs, shanks and feet, and feebly defined, sheath-like, undulating fringe on posterior margins of feet and lower arms (vs obtusely pointed snout in lateral aspect, glandular warty dorsal surfaces on forelimbs, thighs, shanks and feet; and prominent, sheath-like, undulating fringe on posterior margins of feet and lower arms in *P. mittermeieri*). (Figs 1 & 2).

Colour Dorsum light grey-brown with red-brown patch in middle and two black bands extending from upper flanks to mid-back. Dorsal part of head also grey-brown, with symmetrical black marking on occipital. Yellowish-golden patch on shoulder. Anterior thighs have wide, dark brown bands. (Figs 1 & 2). Venter ashy-yellow with black dots.

Habits Can be seen vocalizing while perched on leaf litter and low shrubs about 0.3–2m above the ground at dusk and during the night.

Habitat and Distribution Mainly inhabits leaf litter and low vegetation, and cardamom plantations in thick canopy forests. Restricted to elevations above 1,000m in the Rakwana Mountains (such as the Morningside and Enasalwatta regions of Sinharaja Forest Reserve).

Status Endemic.

IUCN Red List Category Critically Endangered.

Fig. 1 Dorsolateral aspect

Fig. 2 Lateral aspect

DILMAH SHRUB FROG *Pseudophilautus dilmah*

(Dilmah panduru mädiyã)

First described as *Pseudophilautus dilmah* by L. J. M. Wickramasinghe, I. N. Bandara, D. R. Vidanapathirana, K. H. Tennakoon, S. R. Samarakoon and N. Wickramasinghe in 2015.

Size SVL 19–20mm.

Identification Features Small, elongated frog with distinct tympanum and supratympanic fold, rounded canthal edges, medially webbed feet, tuberculated dorsum, horny tubercles on upper eyelids, calcar (spine) at tibiotarsal articulation (heel), less pronounced tarsal fold, and dermal fringes on fingers. Head dorsally convex. Snout bluntly pointed dorsally and ventrally. Internarial space and loreal region concave. Interorbital space convex. Nostrils oval in shape and closer to tip of snout than to eye. Tympanum oval, and smaller than half of eye diameter. Sides of head, upper arms, lower arms, hands and feet smooth. Anterior and posterior dorsum has tubercles and no horny spinules. Upper parts of flanks weakly granular and lower parts strongly granular. Anterior, dorsal and posterior parts of thighs smooth. Throat granular and margin of throat smooth with prominent tubercles. Upper arms granular and lower arms smooth. Chest and belly coarsely granular. Resembles *P. hankeni* and *P. schmarda*. Can be differentiated from *P. hankeni* by rounded snout when viewed laterally and dermal fringes on fingers (vs pointed snout laterally and absence of dermal fringes on fingers in *P. hankeni*). Differs from *P. schmarda* by rounded snout when viewed laterally, and rounded canthus edge (vs obtusely pointed snout and sharp canthus edge in *P. schmarda*). (Figs 1 & 2).

Colour Dorsal aspect cream with light brown patches. Lateral aspect of body has light olive-greenish tinge. Dark brown cross-band between eyes and pair of dark brown dots behind cross-band. Limbs cream, and fingers and toes have light brown cross-bands. Dark brown blotches on groin. (Figs 1 & 2).

Habits Found in submontane and montane forests, and regenerated forests, as well as disturbed areas with no canopy cover. Commonly perches on high bushes about 1–2m above ground level.

Habitat and Distribution Known only from Loolkandura Estate, Deltota, central Sri Lanka, at elevations above 1,300m.

Status Endemic.

IUCN Red List Category Not Evaluated.

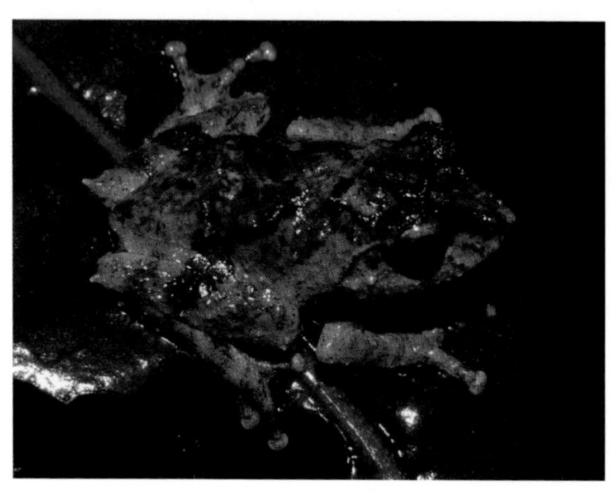

Fig. 1 Dorsolateral aspect

Fig. 2 Lateral aspect, with horny spinules indicated by arrows

DIMBULLA SHRUB FROG *Pseudophilautus dimbullae*

(Dimbulla panduru mädiya)

First described as *Rhacophorus dimbullae* by Benjamin Shreve in 1940. After changes to the generic name (that is, to *Philautus dimbullae*), currently recognized as *Pseudophilautus dimbullae*. Known only from the holotype. There have been no records since its original collection, and it is now believed to be extinct.

Status Endemic.

IUCN Red List Category Extinct.

QUEENWOOD SHRUB FROG *Pseudophilautus eximius*

(Queenwood panduru mädiya)

First described as *Philautus eximius* by Benjamin Shreve in 1940. Currently recognized as *Pseudophilautus eximius*. Known only from the holotype and there have been no records of the species since 1933. It is now believed to be extinct.

Status Endemic.

IUCN Red List Category Extinct.

BLUNT-SNOUTED SHRUB FROG *Pseudophilautus extirpo*

(Mota hombu panduru mädiya)

First described as *Philautus extirpo* by Kelum Manamendra-Arachchi and Rohan Pethiyagoda in 2005 from a specimen in the Natural History Museum, Basel, Switzerland, collected in Sri Lanka in 1882. Recognized as *Pseudophilautus extirpo*. No records since initial collection, and it is now believed to be extinct.

Status Endemic.

IUCN Red List Category Extinct.

Leaf-nesting Shrub Frog *Pseudophilautus femoralis*

(Pala panduru mädiyã)

First described as *Ixalus femoralis* by A. C. L. G. Günther in 1864. After several changes to the generic name (that is, to *Rhacophorus femoralis*, *Philautus femoralis* and *Kirtixalus femoralis*), it is currently known as *Pseudophilautus femoralis*.

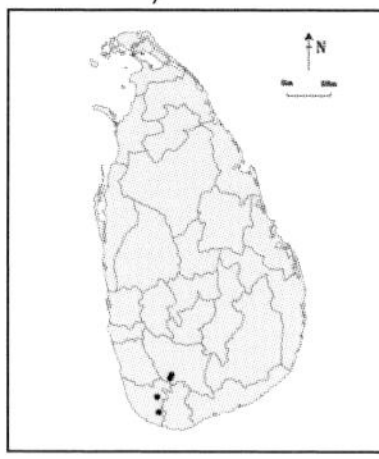

Size SVL 23–30mm.

Identification Features Small, elongated frog with rounded canthal edges, medially webbed feet, finely granular dorsum and granular belly. Lacks distinct tympanum and supratympanic fold. Head longer than broad. Snout rounded and equal to eye diameter. Loreal region concave. Both fingers and toes have supranumerary tubercles. Both edges of digits have cutaneous fringes. Shanks have undulating white or cream-coloured fold of skin. Venter granular. Resembles *P. mooreorum* and *P. poppiae*. Can be differentiated from *P. mooreorum* by dorsally smooth limbs (vs shagreened limbs dorsally). Differs from *P. poppiae* by having head with convex dorsal surface and dorsally smooth limbs (vs having head with flat dorsal surface and dorsally horny limbs in *P. poppiae*). (Figs 1–3).

Colour Dorsum bright green. Some individuals have a few scattered yellow, black or brown spots on dorsal aspect (Figs 1–3). Belly a mixture of light yellow and white or pale pink.

Habits Nocturnal, perching on bushes and low vegetation 1–2m above the ground in forests. Slow-moving species. Females attach their direct-developing transparent eggs to undersurfaces of leaves.

Habitat and Distribution Found in montane cloud forests and occasionally in anthopogenic habitats. Restricted to elevations of 1,200–2,500m in central Sri Lanka, including Horton Plains, Haggala, Ambewela, Agra-Bopath, Sripada and Nuwara Eliya.

Status Endemic.

IUCN Red List Category Endangered.

Fig. 1 Dorsolateral aspect

Fig. 2 *Dorsolateral aspect*

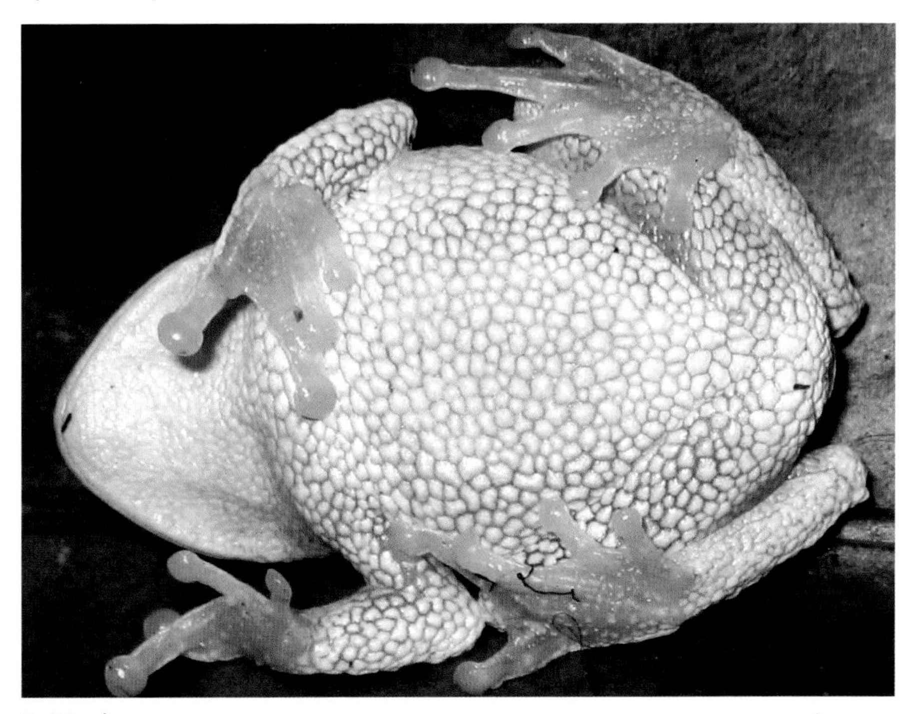

Fig. 3 *Ventral aspect*

Ferguson's Shrub Frog *Pseudophilautus fergusonianus*

(Fergasonge panduru mädiyā)

First described as *Rhacophorus fergusonianus* by Ernst Ahl in 1927. After changes to the generic name (to *Philautus fergusonianus* and *Kirtixalus fergusonianus*), it is currently known as *Pseudophilautus fergusonianus*.

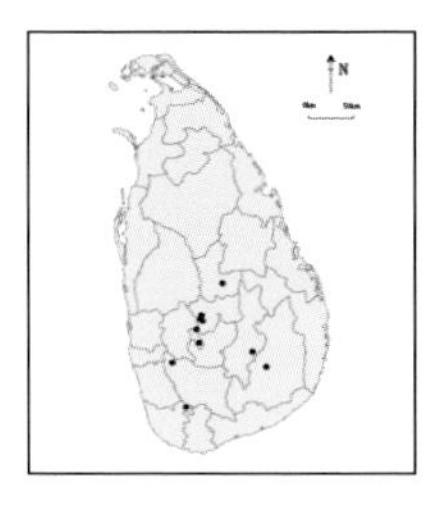

Size SVL 25–45mm.

Identification Features Small, flattened frog with rounded canthal edges, distinct supratympanic fold and tympanum, medially webbed feet, glandular and warty dorsum, and granular belly. Tympanum rounded. Posterior sides of lower arms and feet have row of distinct white glandular tubercles. Head broader or equal to length. Snout short and subtriangular. Loreal region concave. Fingers rudimentarily webbed. Resembles *P. hallidayi*, but can be differentiated by triangular digital discs and shagreened anterior dorsum (vs oval digital discs and glandular-warty anterior dorsum in *P. hallidayi*). (Figs 1 & 2).

Colour Dorsum a mixture of yellow, light brown, tinge of green and ash, with irregular dark markings. Venter pale yellow or brown. Dorsal sides of limbs have brown cross-bars. (Figs 1 & 2). Throat, chin and chest areas have irregular white spots. Belly light yellow with few brown spots.

Habits Arboreal and nocturnal species that can be seen on wet, moss-covered boulders in streams, on tree trunks and moist walls, and in wells. By day can be seen in bathrooms and caves.

Habitat and Distribution Occurs in both forest habitats and anthropogenic habitats such as home gardens and plantations. Found mainly in wet zone at 300–1,000m above sea level. Has been located in Puwakpitiya-Knuckles, Peradeniya, Gampola, Ampitiya, Ambagamuwa, Kuruwita, Udawattekele, Gannoruwa, Hanthana, Monaragala, Pitadeniya-Sinharaja and Lunugala.

Status Endemic.

IUCN Red List Category Least Concern.

Fig. 1 Dorsolateral aspect

Fig. 2 *Lateral aspect*

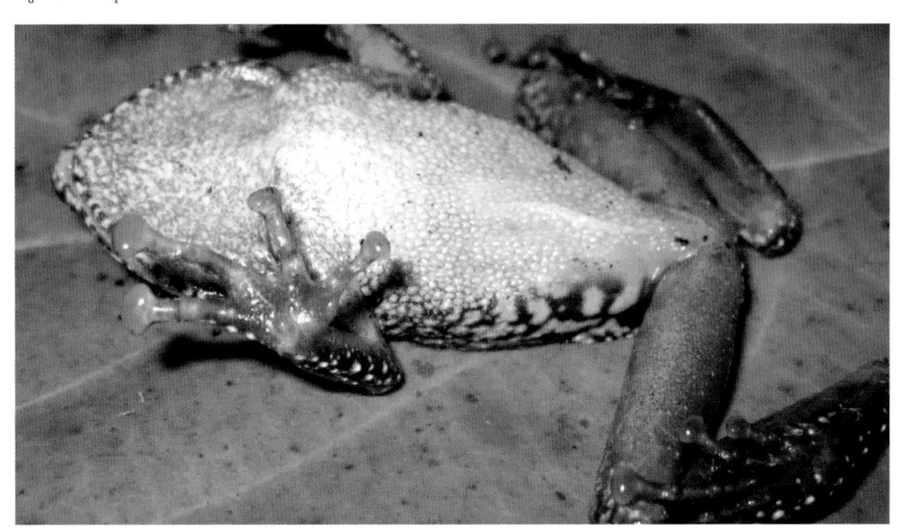

Fig. 3 *Ventral aspect*

Leaf-dwelling Shrub Frog *Pseudophilautus folicola*

(Vakutu-kola panduru mädiyã)

First described in 2005 as *Philautus folicola* by Kelum Manamendra-Arachchi and Rohan Pethiyagoda. Currently recognized under the genus *Pseudophilautus*.

Size SVL 23–30mm.

Identification Features Small, flat, elongated frog with sharp canthal edges, distinct tympanum and supratympanic fold, horny tubercle on each upper eyelid, medially webbed feet, smooth or granular anterior dorsum, smooth posterior dorsum and granular belly. Supernumerary tubercles on palms. Tympanum oval and vertically arranged. Snout and interorbital area smooth. Sides of head smooth, shagreened or with a few scattered glandular warts. Anterior dorsum smooth or granular and posterior dorsum smooth. Flanks granular. Forelimbs and hindlimbs smooth dorsally. Throat and chest granular or smooth. Belly and undersides of thighs granular. Resembles *P. schneideri*, and distinguished by rounded or truncated snout in lateral aspect, concave loreal region, dermal fringes on fingers and granular thighs from underside (vs obtusely pointed snout, flat loreal region and smooth thighs from underside, and absence of dermal fringes on fingers in *P. schneideri*). (Figs 1 & 2). Also resembles *P. stictomerus*.

Colour Dorsum a mixture of dark brown, beige to brick-red. with narrow yellow stripe on mid-dorsum from tip of snout to vent (most individuals do not have stripe). Dorsal sides of limbs have a few brown cross-bars. Forelimbs dark brown with blackish dorsal cross-bars and ventrally ashy dark brown. Dorsal parts of thighs, shanks and feet brown with darker cross-bars. (Figs 1 & 2).

Fig. 1 Dorsolateral aspect

Habits Nocturnal species that hides in leaf litter, underneath logs, and in crevices and tree-holes during the day. Common on branches or leaves 0.3–2m above ground level. Males known to call while staying hidden inside curled leaves.

Habitat and Distribution Habitat generalist that occurs in shaded places, including lowland rainforests, forest edges and anthropogenic habitats at 60–660m above sea level. Recorded from lowland rainforests in Ambagamuwa, Hiyare, Kottawa, Haycock, Gilimale, Induruwa, Udamaliboda and Kanneliya.

Status Endemic.

IUCN Red List Category Vulnerable.

Fig. 2 Lateral aspect

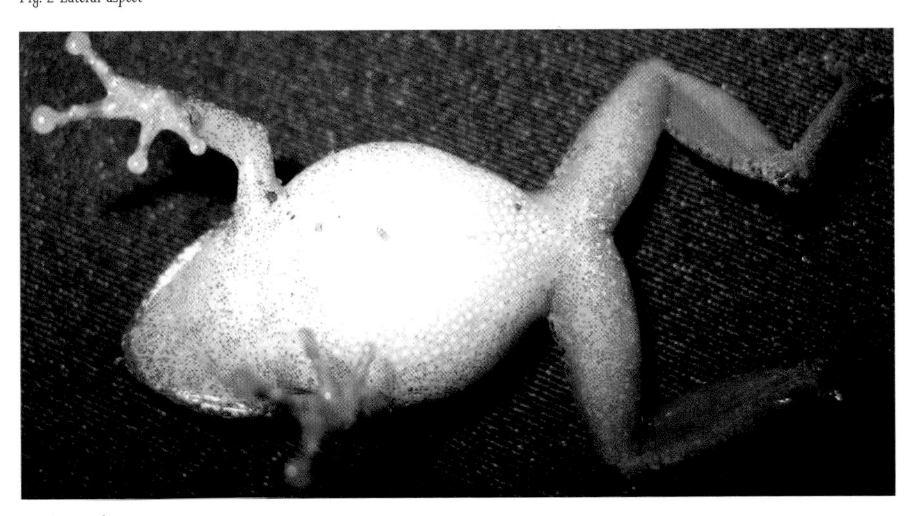

Fig. 3 Ventral aspect

FRANKENBERG'S SHRUB FROG *Pseudophilautus frankenbergi*

(Frankenburge panduru mädiyā)

First described as *Philautus frankenbergi* by Madhava Meegaskumbura and Kelum Manamendra-Arachchi in 2005. Generic name *Philautus* later changed to *Pseudophilautus*, and species is currently documented as *Pseudophilautus frankenbergi*.

Size SVL 26–38mm.

Identification Features Small, stout frog with dorsally flattened head, sharp canthal edges, distinct tympanum and supratympanic fold, horny spinules on dorsum, medially webbed feet, and heavily granular belly and groin. Loreal region concave. Interorbital and internasal spaces flat. Tympanum oval, and vertically arranged. Lateral dermal fringes on fingers. Snout, interorbital space, dorsum and upper flanks shagreened. Sides of head, dorsal parts of forelimbs, and shanks have glandular warts and lower flanks granular. Dorsal parts of thighs and pes smooth. Throat, chest and undersides of thighs granular. Resembles *P. microtympanum* and *P. silus* but can be differentiated by downwards curved supratympanic fold and heavily granular flanks (vs absence of downwards curved supratympanic fold and heavily granular flanks; horny spinules on dorsum and cross-bars on limbs in *P. microtympanum* and *P. silus*). (Figs 1–3).

Colour Comprises a few colour variations. Dorsum ranges from ashy-brown and reddish-brown, to greenish. Some individuals have cream, dark brown or white spots on dorsum. Inguinal zone marbled with black and white. (Figs 1–3).

Habits Nocturnal, ground-nesting species. Hides in leaf litter, underneath decaying logs, in crevices in trees, and among roots in closed canopy forests by day. Can also be seen on shrubs and occasionally on moss- and fern-covered rocks.

Habitat and Distribution Mainly confined to cloud forests in central Sri Lanka, including Piduruthalagala, Horton Plains and Namunukula Peak at above 1,850m.

Status Endemic.

IUCN Red List Category Endangered.

Fig. 1 Dorsolateral aspect

Fig. 2 Dorsolateral aspect

Fig. 3 Lateral aspect

KNUCKLES SHRUB FROG *Pseudophilautus fulvus*

(Dumbara panduru mädiyã)

First described as *Philautus fulvus* by Kelum Manamendra-Arachchi and R. Pethiyagoda in 2005. Due to change of generic name to *Pseudophilautus*, current name is *Pseudophilautus fulvus*.

Size SVL 33–47mm.

Identification Features Small to medium-sized, stout frog with sharp canthal edges, distinct supratympanic fold and tympanum, medially webbed feet, calcar (spur) at tibiotarsal articulation (heel), shagreened dorsum and granular belly. Head dorsally convex. Snout rounded in lateral aspect. Loreal region concave and interorbital space flat. Tympanum oval, and arranged vertically. Yellow oval nuptial pad. Dermal fringes on fingers. Dorsal and lateral parts of snout, sides of head, interorbital area, upper flanks, dorsal areas of forelimbs and thighs shagreened. Shanks and feet smooth. Lower flanks, throat, chest, belly and thighs granular. Resembles *P. silus* and can be differentiated by rounded snout in lateral aspect (vs oval snout in lateral aspect in *P. silus*). (Figs 1 & 2).

Colour Dorsum reddish-brown or brownish-grey with symmetrical dark brown markings. Loreal region, dorsal parts of limbs, thighs and shanks light brown with dark brown cross-bars. Posterior thighs contain light brown patches. Dorsal aspects of feet and webbing dark brown. (Figs 1 & 2).

Habits Nocturnal species that hides in shady places by day. Adults perch on moss-covered rocks, logs or branches 2–4m above the ground at night.

Habitat and Distribution Habitat generalist, found mainly in closed canopy habitats that include dry mixed evergreen forests to submontane forests, anthropogenic habitats, secondary forests and cardamom plantations. Restricted to the Knuckles Mountain Range at 450–1,200m.

Status Endemic.

IUCN Red List Category Endangered.

Fig. 1 Dorsolateral aspect

Fig. 2 *Lateral aspect*

Fig. 3 *Ventral aspect*

Halliday's Shrub Frog *Pseudophilautus hallidayi*

(Halidege panduru mädiyã)

First described as *Philautus hallidayi* by Madhava Meegaskumbura and Kelum Manamendra-Arachchi in 2005. With the change of the generic name, now known as *Pseudophilautus hallidayi*.

Size SVL 33–43mm.

Identification Features Small, flat frog with rounded canthal edges, distinct supratympanic fold and tympanum, medially webbed feet, glandular and warty dorsum and granular belly. Head dorsally convex or flat. Snout rounded in lateral aspect. Loreal region concave. Interorbital space flat and internasal space flat or concave. Tympanum rounded. Dermal fringes on fingers. Distinct glandular warts on snout, interorbital space, sides of head, dorsum, palms, outer edges of lower arms, outer edge of pes, and tibiotarsal articulation. Supernumerary tubercles on palms. Upper flanks have glandular warts and lower flanks granular. (Figs 1 & 2). Belly and undersides of thighs granular. Resembles *P. fergusonianus*, and can be differentiated by oval digital discs and glandular-warty anterior dorsum (vs triangular digital discs and shagreened anterior dorsum in *P. hallidayi*).

Habits Nocturnal species that hides under boulders and rock crevices along streams by day.

Habitat and Distribution Common on boulders along streams in closed canopy forests in wet lowland and submontane regions. Known from a few localities in central Sri Lanka at 500–1,200m, including Hantana Mountain Range, Peradeniya and Tonacombe Estate-Namunukula.

Status Endemic.

IUCN Red List Category Vulnerable.

Fig. 1 Dorsolateral aspect

Fig. 2 Dorsolateral aspect

Fig. 3 Ventral aspect

PATTIPOLA SHRUB FROG *Pseudophilautus halyi*

(Pattipola panduru mädiya)

First described as *Ixalus halyi* by George Boulenger in 1904. After changes to the generic name (that is, to *Rhacophorus halyi* and *Philautus halyi*), currently documented as *Pseudophilautus halyi*. Known only from the holotype, which was collected in Pattipola. There have been no records of it since 1899, and it is now believed to be extinct.

Status Endemic.

IUCN Red List Category Extinct.

HANKEN'S SHRUB FROG *Pseudophilautus hankeni*

(Hankenge panduru mädiyã)

First described as *Pseudophilautus hankeni* by Madhava Meegaskumbura and Kelum Manamendra-Arachchi in 2011. Previously considered as *P. schmarda*.

Size SVL 18–22mm.

Identification Features Small, elongated frog with distinct tympanum and supratympanic fold, rounded canthal edges, rudimentarily webbed feet, tuberculated dorsum, horny tubercles on upper eyelids, calcar (spur) at tibiotarsal articulation (heel), tuberculated dermal fold on outer edges of lower arms and feet, and dermal fringes on fingers. Head laterally convex. Snout rounded in dorsal aspect. Loreal region concave. Interorbital space convex and internarial space concave. Tympanum oval. Dorsal and lateral parts of head, body, and upper parts of flanks have glandular warts bearing horny spinules. Dorsal and lateral parts of upper arms, lower arms, thighs, shanks and feet have glandular warts with horny spinules. Lower parts of flanks, throat, chest and belly granular. Resembles *P. dilmah* and *P. schmarda*. Can be differentiated from *P. dilmah* by pointed snout when viewed laterally and absence of dermal fringes on fingers (vs rounded snout laterally and dermal fringes on fingers in *P. dilmah*). Differs from *P. schmarda* by rounded canthal edges and toes with dermal fringes (vs sharp canthal edges and toes without dermal fringes). (Figs 1 & 2).

Colour Highly variable. Dorsal and lateral parts of head brown to dark green with symmetrical darker patches. Upper areas of eyelids and posterior area of interorbital have distinct black patches. Upper flanks greyish-brown and lower flanks pale yellowish-grey. Loreal region brown. Dorsal and lateral parts of limbs brown. Lower arms, thighs, shanks and feet have brown patches. (Figs 1 & 2).

Habits Nocturnal species that can be seen on leaves of shrubs at 0.3–1.0m above the ground.

Habitat and Distribution Occurs in submontane and montane cloud forests. Can also be seen in cardamom plantations grown within montane forests. Restricted to regions above 1,200m in the Knuckles Mountain Range. Known from Bambarella, Deanstone and Riverstone.

Status Endemic.

IUCN Red List Category Endangered.

Fig. 1 Dorsolateral aspect

Fig. 2 Lateral aspect

Hoffmann's Shrub Frog *Pseudophilautus hoffmanni*

(Hoffmange pañduru mädiyã)

First described as *Philautus* by Kelum Manamendra-Arachchi and Rohan Pethiyagoda in 2005. Since all Sri Lankan *Philautus* species were placed in the genus *Pseudophilautus*, its current name is *Pseudophilautus hoffmanni*. However, it was stated that *P. hoffmanni* would be a junior synonym of *P. asankai* at the upcoming revision of the *Pseudophilautus* genus (Madhava Meegaskumbura, personal communication, IUCN Red List meeting, 2020).

Size 21–25mm.

Identification Features Small frog with stout body, dorsally flat head, rounded snout in lateral aspect, distinct tympanum, indistinct supratympanic fold, rounded canthal edges, flat loreal region, flat interorbital and internasal spaces, and medially webbed toes. Lateral dermal fringes on fingers. Snout, interorbital space, sides of head, dorsum and upper flanks smooth. Lower flanks glandular. Dorsal parts of forelimbs, thighs, shanks and pes smooth. Throat, chest and belly granular with smooth under-thighs. *Pseudophilautus hoffmanni* resembles *P. asankai* and can be distinguished from it by flat interorbital space, distinct tympanum and absence of horn-like spinules on lower flanks (vs interorbital space convex, tympanum indistinct and horn-like spinules on lower flanks in *P. asankai*). (Figs 1–3).

Colour Dorsum and limbs highly variable, with colours ranging from whitish-grey to yellowish-orange and green. Some individuals have darker spots on dorsum and limbs. (Figs 1–3).

Habits Nocturnal species that hides under leaves or on leaf axils, often in well-illuminated habitats. Males vocalize while perched on leaves ~0.3–1m above the ground.

Habitat and Distribution Occurs in closed canopy cloud forests, forest edges and cardamom plantations. Restricted in distribution to elevations above 900m in the Knuckles Mountain Range, and known from Corbett's Gap, Bambarella, Dotalugala, Hunasgiriya and Riverstone.

Status Endemic.

IUCN Red List Category Not Evaluated.

Fig. 1 Dorsolateral aspect

Fig. 2 Dorsolateral aspect

Fig. 3 Lateral aspect

Anthropogenic Shrub Frog *Pseudophilautus hoipolloi*

(Gewathu pañduru mädiyã)

First described as *Philautus hoipolloi* by Kelum Manamendra-Arachchi and Rohan Pethiyagoda in 2005. Since all Sri Lankan *Philautus* species were placed in the genus *Pseudophilautus*, the current name is *Pseudophilautus hoipolloi*. However, it was stated that *P. hoipolloi* would be a junior synonym of *P. pleurotaenia* at the upcoming revision of the *Pseudophilautus* genus (Madhava Meegaskumbura, personal communication, IUCN Red List meeting, 2020).

Size SVL 22–29mm.

Identification Features Small frog with stout body, dorsally convex head, blunt snout in lateral aspect, rounded canthal edges, concave internarial space, flat loreal region and interorbital space, distinct tympanum, indistinct supratympanic fold and medially webbed toes. Distinct, rounded and oblique in shape. Lateral dermal fringes on fingers. Snout, sides of head, interorbital area, and dorsum smooth or shagreened, and dorsal parts of forelimbs, thighs, shanks and feet smooth. Throat and chest shagreened. Belly and under-thighs granular. Resembles *P. pleurotaenia*, and can be distinguished from it by indistinct supratympanic fold and concave internarial region (vs prominent supratympanic fold and flat internarial region in *P. pleurotaenia*). (Figs 1–3).

Colour Dorsum highly variable, ranging from light green to brown. Most common colour phase light green with trace of brown or brownish-green with light brown mottling dorsally. Lateral sides of body have dark brown or black spots and dark mottling on inner and outer surfaces of thighs. (Figs 1–3). Some individuals have light cheek-spot.

Habits Entirely nocturnal. Males perch at 1–3m on shrubs and leaves. Found around forest edges and anthropogenic habitats with low vegetation cover.

Habitat and Distribution Encountered in open habitats such as home gardens and

Fig. 1 Dorsolateral aspect

forest edges. Broadly distributed across lowland rainforests of southwestern wet zone at elevations of 15–684m. Known from Hiyare, Kithulgala, Kottawa, Udugama, Haycock and Yagirala Forest.

Status Endemic.

IUCN Red List Category Not Evaluated.

Fig. 2 Dorsolateral aspect

Fig. 3 Lateral aspect

WEBLESS SHRUB FROG *Pseudophilautus hypomelas*

(Patala rahita pañduru mädiyã)

First described as *Ixalus hypomelas* by A. C. L. G. Günther in 1876. After changes to the generic name (that is, to *Rhacophorus hypomelas* and *Philautus hypomelas*), currently known as *Pseudophilautus hypomelas*. Known only from syntypes deposited in the Natural History Museum, London, until recently rediscovered in Sri Pada Sanduary by Mendis Wickramasinghe and co-workers in 2013.

Size SVL 16–23mm.

Identification Features Small frog with rounded canthal edges, distinct supratympanic fold and tympanum, rudimentarily webbed feet, glandular-warty dorsum and granular belly. Horny spinules can be seen on eyelids and a few are scattered on dorsum. Loreal region concave and snout acuminate in lateral aspect. Can be confused with *P. folicola* but differentiated by acuminate snout in lateral aspect, rounded canthal edges and unique colour pattern (vs oval snout in lateral aspect, sharp canthal edges and unique colour pattern). (Figs 1–3).

Colour Dorsum beige with pair of broad, dark brown, longitudinal dorsal bands extending from back of eye to groin. Dark brown, triangular or 'T'-shaped patch projecting towards vent between eyes. Similarly, dark brown band from tip of snout to front of eye laterally (Figs 1–3). Venter off white with dark brown blotches. In some individuals, venter thickly mottled with brown, with thin cross (+) mark running across entire venter.

Fig. 1 Dorsolateral aspect

Habits Nocturnal species that can be seen in bushes and shrubs from a height of 0.6–1.5m at night.

Habitat and Distribution Found in relatively open areas adjacent to lowland rainforests and submontane forests. So far known from elevations of 600–1,300m in the Sri Pada mountain range.

Status Endemic.

IUCN Red List Category Endangered.

Fig. 2 Dorsolateral aspect

Fig. 3 Lateral aspect

Jagath Gunawardana's Shrub Frog *Pseudophilautus jagathgunawardanai*

(Jagath Gunawardanage pañduru mädiyã)

First described as *Pseudophilautus jagathgunawardanai* by by L. J. M. Wickramasinghe, D. R. Vidanapathirana, M. D. G. Rajeev, S. C. Ariyarathne, A. W. A. Chanaka, L. L. D. Priyantha, I. N. Bandara and N. Wickramasinghe in 2013.

Size SVL 35–41mm.

Identification Features Small, elongated frog with distinct tympanum and supratympanic fold, sharp canthal edges, medially webbed feet, lightly tuberculated dorsum, lingual papilla and dermal fringes on toes. Head dorsally convex. Snout rounded in both ventral and lateral aspects, and truncated in dorsal aspect. Internasal space concave and canthal ridge sharp. Loreal region and interorbital space convex. Nostrils oval and closer to tip of snout than to eye. Supernumerary tubercles on all fingers, on toes, feet and palms. Inner metatarsal tubercle absent. Dorsal and lateral parts of snout and interorbital space smooth. Head laterally smooth. Two cross dermal fringes, one connecting fronts of eyes and the other placed behind them. Faint median dermal ridge from tip of snout to vent. Upper parts of flanks smooth and lower parts weakly granular. Upper arms smooth and forearms weakly tubercular. Inner and outer thighs smooth and dorsally weakly tubercular. Throat, chest, forearms and upper arms weakly granular in ventral aspect. Belly granular. Resembles *P. eximius* but differs by dorsally convex head, convex loreal region and absence of dermal fringes on fingers (vs dorsally flattened head, concave loreal region and dermal fringes on fingers in *P. eximius*). (Figs 1–3).

Colour Dorsally light brown with green tinge on sides. Dark brown circle on snout and broad dark brown band connects upper eyelids. Dark brown, triangular patch pointed towards head on anterior dorsum. Another broad dark brown cross-band on sacrum. Sides light olive-green. Limbs light brown with green tinge. Forelimbs, hindlimbs, fingers and toes have dark brown cross-bands. (Figs 1–3). Throat and chest marbled with brown blotches, and belly, hands, feet and webbing darker in colour.

Habits Common on tree trunks with lichens and on shrubs from the forest floor, to a height of 6m above the ground. Hides in tree bark by day.

Habitat and Distribution Found mainly in cloud forests of the Sri Pada Mountain (Peak Wilderness area) at 1,600–1,750m.

Status Endemic.

IUCN Red List Category Critically Endangered.

Fig. 1 Dorsal aspect

Fig. 2 Dorsolateral aspect

Fig. 3 Dorsal aspect

Karunarathna's Shrub Frog *Pseudophilautus karunarathnai*

(Karunarathnage pañduru mädiyã)

First described as *Pseudophilautus karunarathnai* by L. J. M. Wickramasinghe, D. R. Vidanapathirana, M. D. G. Rajeev, S. C. Ariyarathne, A. W. A. Chanaka, L. L. D. Priyantha, I. N. Bandara and N. Wickramasinghe in 2013.

Size SVL 16–20mm.

Identification Features Very small, elongated frog with distinct tympanum and supratympanic fold, rounded canthal edges, medially webbed feet, horny spinules on head and dermal fringes on fingers and toes. Head dorsally convex. Snout rounded laterally. Loreal region and internasal space concave. Interorbital space convex. Subarticular tubercles prominent and rounded. Nuptial pads present. Supernumerary tubercles absent on fingers, toes and palms. Both inner and outer metatarsal tubercles present. Dorsum smooth anteriorly and spinulated posteriorly. Snout dorsally smooth and laterally weakly tubercular. Lateral head, interorbital space and upper eyelids weakly tubercular. Upper parts of flanks weakly granular and lower parts granular. Upper arms, forearms, hands, inner, outer and dorsal thighs, and feet smooth. Legs and tarsi smooth. Forearms, legs and tarsi smooth, and throat, chest, thighs and belly granular in ventral aspect. Resembles *P. singu*, from which it can be distinguished by smooth anterior dorsum, dorsally smooth thighs and absence of supernumerary tubercles on palms (vs anterior dorsum with scattered glandular tubercles, thighs with dorsally scattered glandular tubercles and supernumerary tubercles on palms in *P. singu*). (Figs 1 & 2).

Colour Dorsal colour uniform cream to beige with dark brown blotches and dark brown cross-band between eyes. Another faint cross-band on snout and broad, dark brown cross-band on sacral hump. Head laterally dark brown with light brown spot (not prominent). Limbs cream coloured, and forelimbs, hindlimbs, fingers and toes have dark brown cross-bands. (Figs 1 & 2).

Habits Found on the forest floor to 1.5m from the ground on shrubs.

Habitat and Distribution Found mainly in lowland rainforests and submontane rainforests, and also in home gardens without much canopy cover. Restricted to the Sri Pada mountain range (Peak Wilderness area) at elevations of 750–1,400m.

Status Endemic.

IUCN Red List Category Critically Endangered.

White-nosed Shrub Frog *Pseudophilautus leucorhinus*

(Sudu nasethi panduru mädiya)

First described as *Ixalus leucorhinus*, and later referred to as *Philautus leucorhinus*. Currently known as *Pseudophilautus leucorhinus*. Known only from type specimen deposited in the Zoological Museum of Berlin, Germany, bearing the number ZMB3057. Recent extensive field surveys have failed to locate it, so it is considered to be extinct.

Status Endemic.

IUCN Red List Category Extinct.

Fig. 1 Dorsolateral aspect

Fig. 2 Dorsal aspect

Haycock Shrub Frog *Pseudophilautus limbus*

(Heycock pañduru mädiyã)

First described as *Philautus limbus* by Kelum Manamendra-Arachchi and Rohan Pethiyagoda in 2005. Currently placed in the *Pseudophilautus* genus, and hence known as *Pseudophilautus limbus.*

Size SVL 25–27mm.

Identification Features Small frog with slender, dorsally concave head, obtusely pointed snout in lateral aspect, rounded canthal edges, concave loreal region, distinct tympanum and supratympanic fold, medially webbed feet, and calcar at tibiotarsal articulation. Interorbital and internarial space concave. Tympanum oval, and vertically arranged. Lingual papilla present. Snout, interorbital area, sides of head and dorsum granular and smooth. Flanks, dorsal parts of forelimbs, thighs, shanks and feet granular. Throat, chest, belly and undersides of thighs granular. (Figs 1 & 2).

Colour Dorsum and dorsal head white with dark brown and reddish-orange patches. Loreal and tympanic region white and brown. Canthal edge and inguinal zone dark brown. Upper half of tympanum dark brown and lower half white. Flank whitish with a few dark brown patches. Upper lip dark brown with white and black patches, and lower lip with some grey patches. Limbs dorsally dark brown and white. (Figs 1 & 2). Belly and ventral aspects of limbs ashy-white.

Habits Type specimen collected from a low (50cm above the ground) branch on border of a forest.

Habitat and Distribution Appears to be restricted to type locality area, including Haycock (Hiniduma), at about 560m elevation.

Status Endemic.

IUCN Red List Category Endangered.

Fig. 1 Dorsolateral aspect

Fig. 2 Dorsolateral aspect

HANDAPAN ELLA SHRUB FROG *Pseudophilautus lunatus*

(Handapana ella pañduru mädiyã)

First described as *Philautus lunatus* by Kelum Manamendra-Arachchi and Rohan Pethiyagoda in 2005. Since the generic name change to *Pseudophilautus*, its current name is *Pseudophilautus lunatus*.

Size SVL 40–41mm.

Identification Features Small frog with distinct tympanum and supratympanic fold, rounded canthal edges, elongated body, dorsally flat head, calcar at tibiotarsal articulation and medially webbed toes. Snout oval in lateral aspect. Loreal region concave. Interorbital and internarial spaces concave. Tympanum oval. Lingual papilla present. Fingers have lateral dermal fringes. Dorsum, snout, interorbital area and sides of head smooth. Upper flanks have glandular warts and lower flank granular. Dorsal parts of forelimbs and thighs smooth. Shanks and feet smooth with a few scattered glandular warts. Throat, chest, belly and under-thighs smooth. Resembles *P. microtympanum* and can be distinguished from it by lingual papilla, concave interorbital space and smooth venter (vs absence of lingual papilla, flat interorbital space and granualar venter in *P. microtympanum*). (Figs 1 & 2).

Colour Mid-dorsum reddish-brown with two dark brown bands on dorsum. Dorsolateral area ashy-brown with dark brown patches. Interorbital bar dark brown with black spots. Flank pale ashy-brown with dark brown patches. Canthal edges, lower eyes, lower area of supratympanic fold and tympanum dark brown. Limbs dorsally ashy-brown with red and dark brown cross-bars, and ventrally dark brown with ashy spots. (Figs 1 & 2). Chin and chest white with dark brown patches.

Habits Nocturnal species that occurs about 1m above ground level at night.

Habitat and Distribution Restricted to elevations above 1,000m in the Rakwana Mountains.

Status Endemic.

IUCN Red List Category Critically Endangered.

Fig. 1 *Dorsolateral aspect*

Fig. 2 *Lateral aspect*

BIGFOOT SHRUB FROG *Pseudophilautus macropus*

(Visala padethi panduru mädiyã)

First described as *Ixalus macropus* by A. C. L. G. Günther in 1869. After changes to the generic name (that is, to *Rhacophorus macropus* and *Philautus macropus*), current name is *Pseudophilautus macropus*.

Size SVL 27–43mm.

Identification Features Small frog with stout to dorsoventrally flattened body, distinct tympanum and supratympanic fold, rounded canthal edges, concave loreal region, lateral dermal fringes on fingers and fully webbed toes. Interorbital and internarial spaces flat. Tympanum oval shaped. Dorsal and lateral parts of snout smooth and interorbital area smooth with a few scattered glandular warts. Dorsum shagreened with a few scattered glandular warts, and sides of head have glandular warts. Lower flanks granular and upper flanks granular with a few scattered glandular warts. Dorsal parts of forelimbs shagreened with a few scattered glandular warts. Dorsal surfaces of thighs, shanks and feet smooth with a few scattered glandular warts. Lingual papilla present. Belly granular. Resembles *P. sarasinorum* and *P. sordidus*. Differs from *P. sarasinorum* by concave internarial space and absence of lateral dermal fringes on fingers (vs flat internarial space and presence of lateral dermal fringes on fingers in *P. sarasinorum*). Distinguished from *P. sordidus* by fully webbed toes (vs medially webbed toes in *P. sordidus*). (Figs 1–3).

Colour Dorsum pale brownish-yellow to brown with two dark brown dorsolateral bands. Interorbital bar and upper flank dark brown. Limbs pale ashy-brown with dark brown dorsal cross-bars. (Figs 1–3). Abdomen pale yellowish-white.

Habits Habitat specialist found on wet boulders, and in rock crevices and tree-holes around streams. Occasionally seen on tree trunks adjacent to streams.

Habitat and Distribution Found in both closed and open canopy habitats close to streams or on boulders in streams. Restricted to the Knuckles Mountain Range in central Sri Lanka at elevations of 600–1,300m above sea level.

Status Endemic.

IUCN Red List Category Vulnerable.

Fig. 1 Dorsolateral aspect

Fig. 2 Dorsolateral aspect

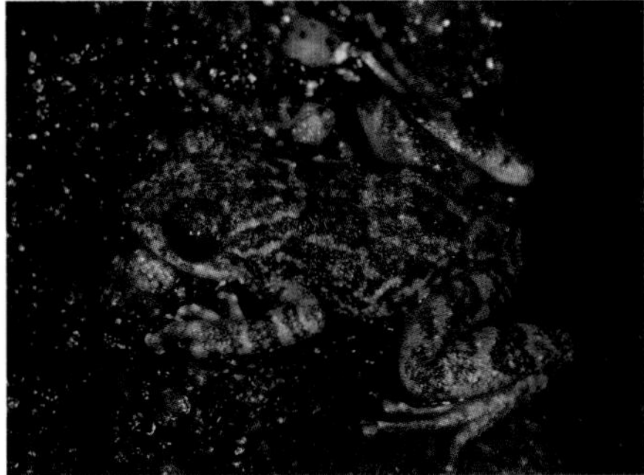

Fig. 3 Lateral aspect

GOOD MOTHER SHRUB FROG *Pseudophilautus maia*

(Sinhala: Mathru paduru mediya)

First described as *Philautus maia* by Madhava Meegaskumbura and co-workers in 2007. Following the generic name revision in 2009, the currently accepted name is *Pseudophilautus maia*. Known from only two specimens collected in 1876 or earlier from Ramboda, Sri Lanka, which were deposited in the Natural History Musum, London, bearing the museum number BMNH 1947.2.7.96. Field surveys in the type locality and other areas during 1993–2003 have failed to relocate the species, so it is now considered extinct.

Status Endemic.

IUCN Red List Category Extinct.

MALCOLMSMITH'S SHRUB FROG *Pseudophilautus malcolmsmithi*

(Malcamsmithge panduru mädiya)

First described as *Rhacophorus malcolmsmithi* by Ernst Ahl in 1927. After several generic name changes (for example, to *Philautus malcolmsmithi*), the species is currently known as *Pseudophilautus malcolmsmithi*. Known only from the type specimen deposited in the Zoological Museum of Berlin, Germany, bearing the museum number ZMB9037. There have been no sightings of the species for more than 70 years, so it is considered extinct.

Status Endemic.

IUCN Red List Category Extinct.

SMALL-EARED SHRUB FROG *Pseudophilautus microtympanum*

(Kudukan pañduru mädiyã)

First described as *Polypedates microtympanum* by Albert Günther in 1859. After changes to the generic name (to *Rhacophorus microtympanum* and *Kirtixalus microtympanum*), it is currently known as *Pseudophilautus microtympanum*.

Size SVL 28–51mm.

Identification Features Small frog with distinct tympanum, supratympanic fold, rounded canthal edges, broad head and concave loreal region medially webbed. All digits contain supernumerary tubercles. Fingers contain dermal fringes. Ventral aspects of thighs roughly granular. Resembles *P. steineri* and can be distinguished by dark brown posterior margins on thighs and pale brown blotches on posterior margins of thighs (vs pale brown posterior margins on thighs and absence of pale brown blotches on posterior margins of thighs in *P. steineri*). (Figs 1–3).

Colour Background colour of dorsum highly variable. Common dorsal colour light brown or grey with symmetrically arranged dark brown or black bands. Both loreal and temporal regions dark brown. Tinge of green on flanks in some individuals. Dorsal sides of limbs bear a few dark cross-bars. (Figs 1–3). Ventral aspect a mixture of pale brown and light purple.

Habits Nocturnal, and hides in various microhabitats, for example underneath logs and debris, inside tree-holes (even about 10m above the ground), in root crevices, under rocks and leaf litter, around bases and roots of grasses, and even inside holes and cracks in the ground in marshes by day or during the dry season. Calls throughout the day and night. Eggs laid (~45 eggs in clutch) in humus or at bases of grass tussocks.

Fig. 1 Dorsolateral aspect

Habitat and Distribution Inhabits all kinds of vegetation types, including anthropogenic habitats, plantations, grassland and forests. Restricted to elevations above 1,500m of the Central Hills. Commonly seen in Horton Plains National Park, Ohiya, Pattipola, Ambewela, Nuwara Eliya, Sri Pada, Loolecondera and Blackpool.

Status Endemic.

IUCN Red List Category Endangered.

Fig. 2 Dorsolateral aspect

Fig. 3 Dorsolateral aspect

Mittermeier's Shrub Frog *Pseudophilautus mittermeieri*

(Mittermeierge pañduru mädiyã)

First described as *Philautus mittermeieri* by Madhava Meegaskumbura and Kelum Manamendra-Arachchi in 2005. Since the generic revision the current name has been *Pseudophilautus mittermeieri*.

Size SVL 18–20mm.

Identification Features Small frog with distinct tympanum and supratympanic fold, dermal fringes on margins of lower arms and tarsi (tarsal fold), rounded canthal edges, medially webbed feet, tuberculated dorsum, prominent calcar (spur) at tibiotarsal articulation (heel) and dermal fringes on fingers. Resembles *P. decoris*, but differs from it by obtusely pointed snout in lateral aspect; glandular warty dorsal surfaces on forelimbs, thighs, shanks and feet; and prominent sheath-like, undulating fringes on posterior margins of feet and lower arms (vs sharp pointed snout in lateral aspect, smooth dorsal surfaces on forelimbs, thighs, shanks and feet, and feebly defined, sheath-like, undulating fringes on posterior margins of feet and lower arms in *P. decoris*). (Figs 1 & 2).

Colour Dorsum dark ashy-olive with orange tubercles on head. Sides of head ashy-brown. Orange pigments at the back. Supratympanic fold orangish-light brown. Dark brown 'X' mark from eyes to mid-back. Posterior thighs and tibias dark brown and anterior thighs contain dark brown cross markings. Lower arm has four, thigh three and shank three dark ashy-brown cross-bars. (Figs 1 & 2).

Habits Nocturnal species that inhabits lowland rainforests and is active on low vegetation at night.

Habitat and Distribution Recorded only from lowland rainforest areas at 60–600m, including in Hiyare, Kanneliya, Kottawa and Beraliya (Elpitiya) forest.

Status Endemic.

IUCN Red List Category Vulnerable.

Fig. 1 Dorsolateral aspect

Fig. 2 *Lateral aspect*

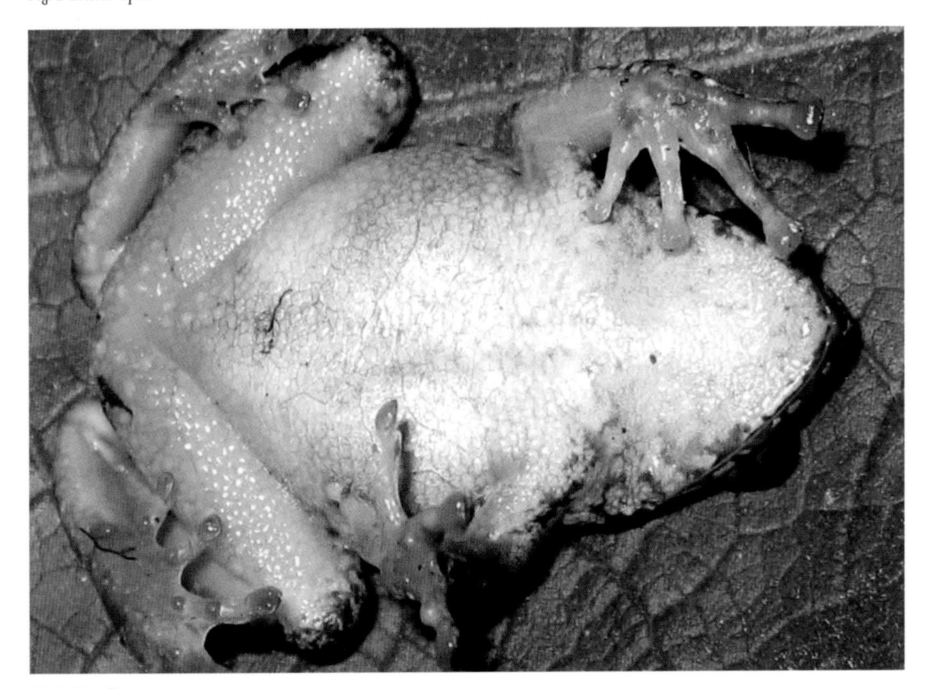

Fig. 3 *Ventral aspect*

Moore's Shrub Frog *Pseudophilautus mooreorum*

(Murige panduru mädiyã)

First described as *Philautus mooreorum* by Madhava Meegaskumbura and Kelum Manamendra-Arachchi in 2005. Since the generic reassignment to *Pseudophilautus*, the current name is *Pseudophilautus mooreorum*.

Size 29–35mm.

Identification Features Small, light greenish, elongated frog with stout, dorsally flat head, rounded canthal edges, medially webbed feet, finely granular dorsum and granular belly. Lacks a distinct tympanum but bears a lightly visible supratympanic fold. Lower flanks smooth. Dorsal parts of forelimbs, thighs, shanks and feet shagreened. Throat, chest, belly and undersides of thighs granular. Dorsum finely granular or shagreened in female. Resembles *P. femoralis* and *P. poppiae*. Can be differentiated from *P. femoralis* by dorsally shagreened limbs and supratympanic fold (vs by dorsally smooth limbs and absence of supratympanic fold in *P. femoralis*). Differs from *P. poppiae* by dorsally flat head and dorsally shagreened limbs (vs dorsally convex head and dorsally horny and spinulated limbs in *P. poppiae*). (Figs 1 & 2).

Colour Dorsum yellowish-green. Upper flanks yellow and white, and lower flanks white. Limbs dorsally green. Discs, venter, dorsal areas of upper arms, outer edges of lower arms, and pes white. Fingers and toes dorsally white or pale green. (Figs 1 & 2). Ventral colour yellowish-white.

Habits Arboreal species that inhabits the forest subcanopy and shrubs in the understorey of forests.

Habitat and Distribution Restricted to montane cloud forests at 1,000–1,800m above sea level in the Knuckles Mountain Range. Recorded in the Knuckles Peak in Bambarella, Riverstone, Corbet's Gap and Deanstone and Hunnasgiriya areas of the Knuckles hills.

Status Endemic.

IUCN Red List Category Critically Endangered.

Fig. 1 Dorsolateral aspect

Fig. 2 *Dorsolateral aspect*

SOUTHERN SHRUB FROG *Pseudophilautus nanus*

(Dakunudiga panduru mädiya)

First described as *Polypedates nanus* by Albert Günther in 1869. After several changes to the generic name (that is, to *Rhacophorus nanus*, *Kirtixalus nanus* and *Philautus nanus*), currently known as *Pseudophilautus nanus*. Known only from the lectotype deposited in the Natural History Museum, London, bearing the museum number 1947.2.7.78. There have been no sightings of the species since the description, and it is now considered to be extinct.

Status Endemic.

IUCN Red List Category Extinct.

POINTED-SNOUT SHRUB FROG *Pseudophilautus nasutus*

(Ulhombu panduru mädiya)

First described as *Ixalus nasutus* by Albert Günther in 1869. After changes to the generic name (that is, to *Rhacophorus nasutus* and *Philautus nasutus*), it is currently known as *Pseudophilautus nasutus*. Known only from the holotype deposited in the Natural History Museum, London, bearing the museum number 1947.2.6.21. There have been no sightings of the species since it was originally described in 1869 despite extensive surveys in Sri Lanka. It is now considered to be extinct.

Status Endemic.

IUCN Red List Category Extinct.

WHISTLING SHRUB FROG *Pseudophilautus nemus*

(Uruhanbana pañduru mädiyã)

First described as *Philautus nemus* by Kelum Manamendra-Arachchi and Rohan Pethiyagoda in 2005. Since the assignment of all Sri Lankan *Philautus* species to *Pseudophilautus*, its current name has been *Pseudophilautus nemus*.

Size SVL 20–24mm.

Identification Features Small frog with slender body, dorsally convex head, distinct tympanum and supratympanic fold, rounded canthal edges, medially webbed toes and obtusely pointed snout in lateral aspect. Loreal region concave. Both interorbital and internarial spaces flat. Tympanum oval and vertically arranged. Snout, sides of head and dorsum have glandular warts. Internarial area, interorbital area and upper flanks smooth. Lower flanks granular. Dorsal aspects of forelimbs and shanks have glandular warts. Dorsal aspects of thighs and feet smooth. Throat and undersides of thighs granular and not rough. Chest and belly granular and rough. Resembles *P. silvaticus*, *P. rus* and *P. singu*. Can be distinguished from *P. silvaticus* and *P. rus* by dermal ridges on dorsum of *P. nemus* (vs absence of dermal ridges on dorsum of *P. silvaticus* and *P. rus*). Can be distinguished from *P. singu* by feebly defined horny spines on upper eyelid (vs prominent horny spines on upper eyelid in *P. singu*). (Figs 1 & 2).

Colour Dorsum reddish-brown with black patches. Interorbital bar black or dark brown. Limbs, fingers and toes have brown cross-bars. Tips of tubercles red and venter pale yellow with black pigments. (Figs 1 & 2).

Habits Usually perches on wet rocks and vegetation in closed canopy rainforests and secondary forests.

Habitat and Distribution Inhabits undisturbed rainforests as well as secondary forests at 50–660m above sea level. Has been seen in a few rainforest habitats, including Beraliyakele, Dellawa, Ambagamuwa and the type locality, Haycock Hill (Hiniduma Forest Reserve).

Status Endemic.

IUCN Red List Category Endangered.

Fig. 1 Dorsolateral aspect

Fig. 2 Lateral aspect

Newton Jayawardane's Shrub Frog *Pseudophilautus newtonjayawardanei*

(Newton Jayawardanage pañduru mädiyā)

First described as *Pseudophilautus newtonjayawardanei* by L. J. M. Wickramasinghe, D. R. Vidanapathirana, M. D. G. Rajeev, S. C. Ariyarathne, A. W. A. Chanaka, L. L. D. Priyantha, I. N. Bandara and N. Wickramasinghe in 2013.

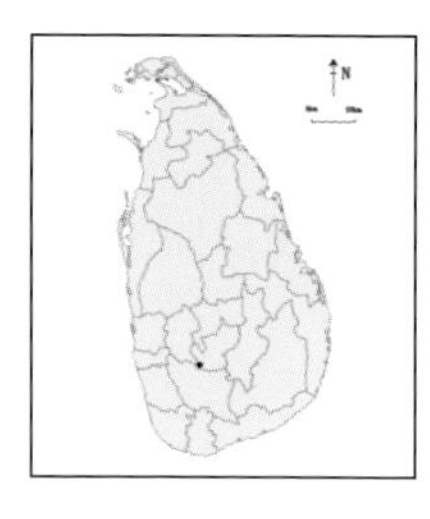

Size SVL 20–24mm.

Identification Features Small frog with elongated body, indistinct tympanum, distinct supratympanic fold, rounded canthal edges, dorsally concave head, rounded snout in lateral, dorsal and ventral aspects, and medially webbed feet. Internasal space, interorbital space and loreal region concave. Dermal fringe on insides of all digits. Supernumerary tubercles on palms and on all fingers. Both dorsal and lateral parts of snout shagreened. Interorbital space shagreened, with prominent ridge. Upper eyelids and head laterally shagreened, with prominent tubercles. Median dermal ridge extends from tip of snout to vent. Dorsum, upper arms, forearms, hands and upper parts of flanks shagreened. Lower parts of flanks weakly granular. Inner thighs smooth and outer thighs weakly granular. Tarsi and feet shagreened. Ventral aspect of throat, chest, belly and upper arms granular. Thighs granular and legs smooth ventrally. Resembles *P. adspersus*, and can be differentiated by rounded canthal edges, laterally rounded snout and inner metatarsal tubercle, and lack of white spots on dorsum (vs sharp canthal edges, laterally truncated snout and white spots on dorsum, and absence of inner metatarsal tubercle in *P. adspersus*). (Figs 1 & 2). Also resembles *P. microtympanum*.

Colour Dorsum bronze with dark brown patches. Off-white vertebral stripe with pair of broad, dark brown longitudinal dorsal bands extends from back of eye to groin. Dark brown cross-band between eyes, and small bronze blotch on snout. Dark brown band on canthal edge. Lateral sides lighter with green tinge, and dark brown spots marbled with off-white spots. Limbs dorsally bronze. Forelimbs and hindlimbs have brown cross-bands (Figs 1 & 2).

Habits Very rare species that perches on leaves about 8m above ground level.

Habitat and Distribution Found in the forest canopy at very high altitudes of 1,500–2,000m in the Sri Pada (Peak Wilderness) mountain range.

Status Endemic.

IUCN Red List Category Critically Endangered.

Fig. 1 Dorsal aspect

Fig. 2 Dorsalateral aspect

Golden-eye Shrub Frog *Pseudophilautus ocularis*

(Ranwan esethi pañduru mädiyã)

First described as *Philautus ocularis* by Kelum Manamendra-Arachchi and Rohan Pethiyagoda in 2005. Since the inclusion of all Sri Lankan *Philautus* species in *Pseudophilautus*, its name has been *Pseudophilautus ocularis*.

Size SVL 23–33mm.

Identification Features Small frog with stout body, dorsally flat head, distinct tympanum and supratympanic fold, rounded canthal edges and medially webbed toes. Snout blunt in lateral aspect. Interorbital space flat and both loreal region and internarial space concave. Tympanum oval. Lateral dermal fringes on fingers. Snout, interorbital area, sides of head, dorsum and upper flanks shagreened or with glandular warts. Lower flanks, and dorsal sides of forelimbs, thighs, shanks and feet have glandular warts. Throat, chest, belly and undersides of thighs granular or with glandular folds. Resembles *P. viridis* and *P. stuarti*. (Figs 1 & 2).

Colour Dorsum and lateral aspect of head and body changeable from light green to dark (blackish) green with a few scattered black spots. Upper flanks dark green and lower flanks yellowish to light green. Inguinal zone and thighs bluish-green. Dorsal sides of both upper and lower arms green. (Figs 1 & 2). Area around vent yellow.

Habits Arboreal species found on trees at night.

Habitat and Distribution Found in closed canopy cloud forests, cardamom plantations within cloud forests and forest edges. Restricted to elevations above 1,000m in the Rakwana mountain range. Known from Morningside, Handapan Ella Plains, Gongala and Enasalwatta in Sinharaja Forest.

Status Endemic.

IUCN Red List Category Critically Endangered.

Sharp-snouted Shrub Frog *Pseudophilautus oxyrhynchus*

(Thiyunu hombu panduru mädiya)

First described as *Ixalus oxyrhynchus* by Albert Günther in 1872. After changes to the generic name (that is, to *Rhacophorus oxyrhynchus* and *Philautus oxyrhynchus*), it is currently recognized as *Pseudophilautus oxyrhynchus* after the generic reassignment to *Pseduophilautus*. Known only from the lectotype deposited in the Natural History Museum, London, bearing the museum number BMNH.1947.2.6.40. There have been no sightings of the species since it was described, and it is now considered to be extinct.

Status Endemic.

IUCN Red List Category Extinct.

Fig. 1 Dorsal aspect

Fig. 2 Lateral aspect

PAPILIATED SHRUB FROG *Pseudophilautus papillosus*

(Dive getiththi pañduru mädiyã)

First described as *Philautus papillosus* by Kelum Manamendra-Arachchi and Pethiyagoda in 2005. Since Sri Lankan and Indian *Philautus* species were placed in the genus *Pseudophilautus*, the current name has been *Pseudophilautus papillosus*. It was stated that *P. papillosus* would be a junior synonym of *P. reticulatus* at the upcoming revision of the *Pseudophilautus* genus (Madhava Meegaskumbura, personal communication, IUCN Red List meeting, 2020).

Size 42–55mm.

Identification Features Medium-sized, elongated frog with dorsally convex or flat head, truncated or oval snout in lateral aspect, sharp canthal edges, concave loreal region, distinct tympanum and supratympanic fold, two short spines (that is, tarsal tubercle and calcar) at heels, fingers with dermal fringes, fully webbed toes and reticulated pattern on posterior thighs. Interorbital and internarial spaces convex or concave. Tympanum oval shaped. Rounded lingual papilla. Dorsal surface finely tuberculated. Chin smooth and chest granular. Ventral aspects of upper arms, abdomen, and ventral and posterior thighs rough and granular. Dorsum, snout, interorbital space, sides of head and upper parts of flanks shagreened with a few scattered glandular warts. Lower flanks granular. Dorsal parts of forelimbs shagreened. Thighs, shanks and feet smooth. Belly and undersides of thighs granular. Resembles *P. reticulatus* and *P. maia*. Can be distinguished from *P. maia* by absence of both tarsal tubercle and calcars at heels in *P. maia*. Can be distinguished from *P. reticulatus* by presence of larger lingual papilla in *P. papillosus* (vs smaller lingual papilla in *P. reticulatus*). (Fig. 1).

Colour Dorsum and dorsal side of head reddish-brown with indistinct dark brown patches (Figs 1). Limbs have indistinct dark brown cross-bars. Chin, chest and belly ashy-brown with dark brown patches (Fig. 1). Flanks ashy-brown and marbled with dark brown. Anterior and posterior thighs ashy-brown, and posterior thighs marbled/reticulated with dark brown (Fig. 2).

Habits Found on branches about 2m above ground level.

Habitat and Distribution Known only from elevations above 1,000m in the Rakwana mountain range, including Handapan Ella Plains and Morningside in Sinharaja.

Status Endemic.

IUCN Red List Category Not Evaluated.

LEOPARD SHRUB FROG *Pseudophilautus pardus*

(Divi panduru mediya)

First described as *Philautus pardus* by Kelum Meegaskumbura and colleagues in 2007. Since all Sri Lankan and Indian *Philautus* species were placed in the genus *Pseudophilautus*, its current name has been *Pseudophilautus pardus*. Known only from the holotype deposited in the Natural History Museum, London, bearing the museum number 1947.2.7.96. There have been no sightings of the species since it was described in 1859, and it is now considered to be extinct. It was stated that *P. pardus* would be a junior synonym of *P. viridis* at the upcoming revision of the *Pseudophilautus* genus (Madhava Meegaskumbura, personal communication, IUCN Red List meeting, 2020).

Status Endemic.

IUCN Red List Category Not Evaluated.

Fig. 1 Lateral aspect

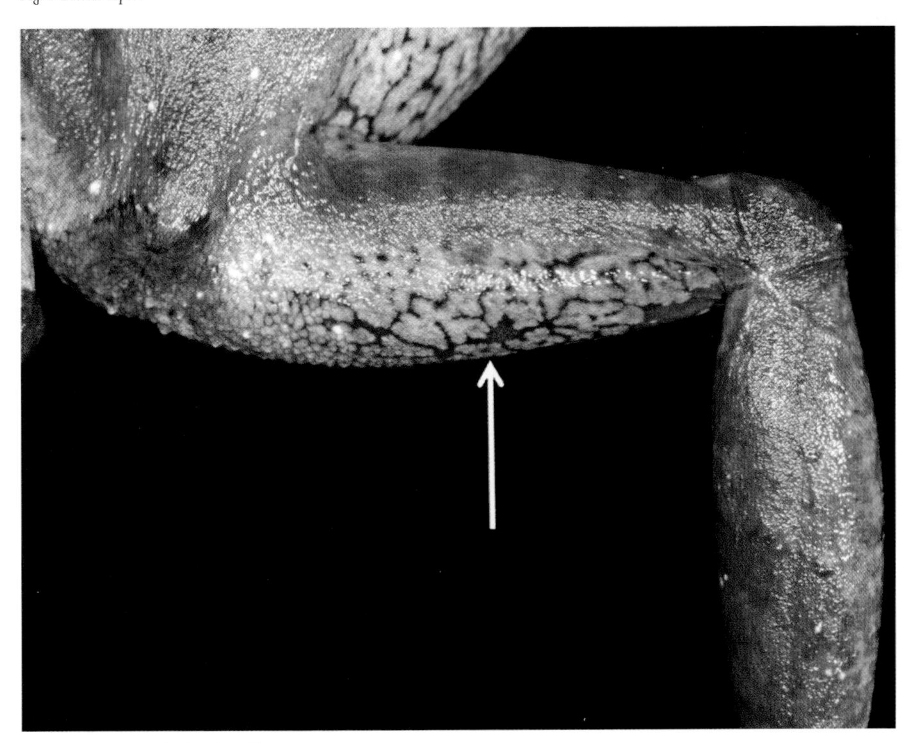

Fig. 2 Reticulation on thigh indicated by arrow

SIDE-STRIPED SHRUB FROG *Pseudophilautus pleurotaenia*

(Pethi thiru pañduru mädiyã)

First described as *Rhacophorus pleurotaenia* by George Albert Boulenger in 1904. After changes to the generic name (to *Kirtixalus pleurotaenia* and *Philautus pleurotaenia*), it is currently known as *Pseudophilautus pleurotaenia*.

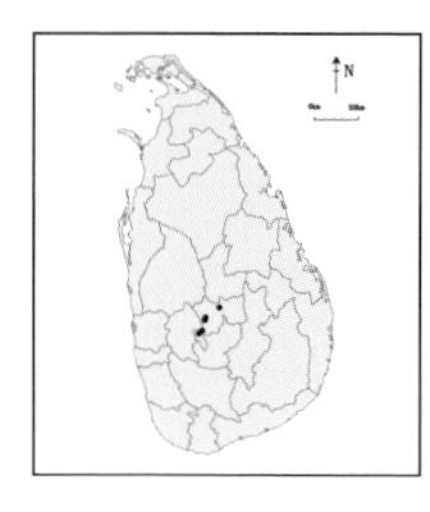

Size SVL 25–30mm.

Identification Features Small, slender-bodied frog with dorsally convex head, truncated snout in lateral aspect, rounded canthal edges, distinct tympanum, fully webbed toes and feebly distinct supratympanic ridge. Interorbital space convex, and both loreal region and internarial space concave. Tympanum half circle, and horizontally arranged. Toes two, three, and four fully webbed. Lateral dermal fringes on fingers. Dorsum shagreened with a few scattered glandular warts and horny spinules. Dorsal and lateral aspects of snout, interorbital area and sides of head shagreened. Upper flanks shagreened. Lower flanks granular. Dorsal parts of forelimbs, thighs, shanks and feet smooth. Throat, chest, belly and undersides of thighs granular. Resembles *P. hoipolloi* and distinguished from it by prominent supratympanic fold, fully webbed toes and flat internarial region (vs indistinct supratympanic fold, medially webbed toes and concave internarial region in *P. hoipolloi*). (Figs 1–3).

Colour Dorsum highly variable, from light greenish-yellow to dark brown. Some individuals may have yellow or white lateral stripes running from back of eye to groin. Black dots may or may not be scattered over body. (Figs 1–3).

Habits Arboreal species that can be seen perched on low vegetation about 1m above the ground. Males usually become active with slight rains, and start calling from shrubs.

Habitat and Distribution Found in forest edges and densely wooded home gardens. Distributed in wet zone of Sri Lanka at 500–800m above sea level. Reported from localities in Sri Lanka such as Kandy, Gannoruwa, Nawalapitiya and Ambagamuwa.

Status Endemic.

IUCN Red List Category Vulnerable.

Fig. 1 Dorsolateral aspect, with tympanum half-circle indicated by arrow

Fig. 2 Dorsolateral aspect

Fig. 3 Lateral aspect

POPPY'S SHRUB FROG *Pseudophilautus poppiae*

(Popige pañduru mädiyã)

First described as *Philautus poppiae* by Madhava Meegaskumbura and Kelum Manamendra-Arachchi in 2005. Following the generic assignment to *Pseudophilautus*, its current name has been *Pseudophilautus poppiae*.

Size SVL 21–26mm.

Identification Features Small, elongated frog with rounded canthal edges, medially webbed feet, finely granular dorsum, feebly distinct supratympanic fold, lateral dermal fringes on fingers and granular belly. Lacks a distinct tympanum. Head dorsally convex, snout rounded in lateral aspect. Interorbital space flat and both loreal region and internasal space concave. Dorsum, snout, interorbital space, sides of head, upper flanks, dorsal areas of forelimbs, thighs, shanks and pes have horny spinules (in male). Lower flanks smooth. Throat, chest and belly granular. Undersides of thighs smooth. Dorsum finely granular or shagreened in female. Resembles *P. femoralis* and *P. mooreorum*. Can be distinguished from *P. femoralis* by dorsally horny, spinulated limbs (vs dorsally smooth limbs in *P. femoralis*). Can be distinguished from *P. mooreorum* by dorsally horny and spinulated limbs (vs dorsally shagreened limbs in *P. mooreorum*). (Figs 1 & 2).

Colour Dorsal and lateral aspects of head and dorsum bright yellowish-green with black dots. Some individuals may have yellow or red spots scattered on dorsum. Inguinal zone and anterior thighs brownish-yellow. Upper arms dorsally greenish-yellow. Outer edges of lower arms have longitudinal white band. Flanks as well as inner sides of upper and lower arms light greenish-yellow. (Figs 1 & 2). Venter pale yellow.

Habits Nocturnal species that can be seen on shrubs and trees.

Habitat and Distribution Found in submontane forests, secondary forests and cardomom plantations. Occasionally seen in tea estates bordering forests. Restricted to elevations above 1,000m in the Rakwana mountain massif. Known from Handapan Ella Plains, Morningside-Sinharaja and Enasalwatta-Sinharaja.

Status Endemic.

IUCN Red List Category Critically Endangered.

Fig. 1 Dorsolateral aspect

Fig. 2 Lateral aspect

Common Shrub Frog *Pseudophilautus popularis*

(Sulaba pañduru mädiyã)

First described as *Philautus popularis* by Kelum Manamendra-Arachchi and Rohan Pethiyagoda in 2005. Since all Sri Lankan and Indian *Philautus* species were placed in the genus *Pseudophilautus*, its current name is *Pseudophilautus popularis*.

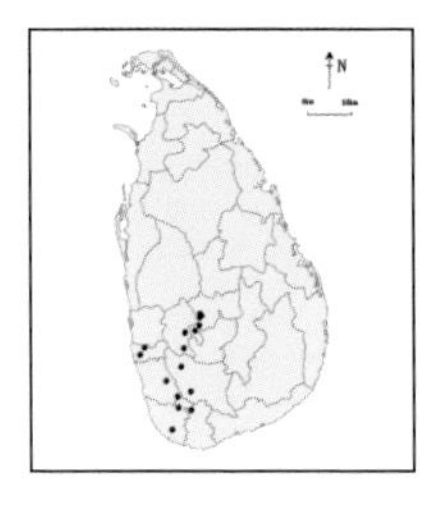

Size SVL 17–25mm.

Identification Features Small frog with dorsally convex head, rounded snout in lateral aspect, rounded canthal edges, concave loreal region, medially webbed toes and distinct tympanum and supratympanic fold. Both interorbital and internarial spaces flat. Tympanum oval shaped and oblique. Supernumarary tubercles on palms and feet. Snout and lateral sides of head have glandular warts. Interorbital area smooth. Dorsum and upper flanks have horn-like spinules. Lower flanks granular. Dorsal parts of forelimbs, thighs and shanks have glandular warts. Feet dorsally smooth. Throat, chest, belly and undersides of thighs granular. Nuptial pad in male. Resembles *P. regius* and can be distinguished by glandular warts on snout and sides of head, and smooth interorbital area (vs horny spinules on snout, sides of head and interorbital area in *P. regius*). (Figs 1 & 2).

Colour Dorsum varies from beige to ashy-light brown. Some individuals have horseshoe-shaped mark on back. Limbs have indistinct brown cross-bars in dorsal aspect. Throat yellowish with brown patches. Belly pale yellow and ventral sides of thighs pale brown. Dorsal aspects of tibias have line of orange-tipped tubercles (Fig. 2).

Habits Common species that perches on low vegetation. Males become active with slight rains and start calling from low vegetations about 0.3–1m above the ground.

Habitat and Distribution Occurs in forest edges, grassland and low vegetation in anthropogenic habitats and along roadsides. Confined to low-country wet zone at 15–1,000m above sea level. Reported from Gampola, Peradeniya, Kandy, Pilimathalawa, Galle, Kelaniya, Kottawa, Deniyaya, Kanneliya, Kudawa, Mathugama, Nawalapitiya, Ambagamuwa and Kithulgala.

Status Endemic.

IUCN Red List Category Least Concern.

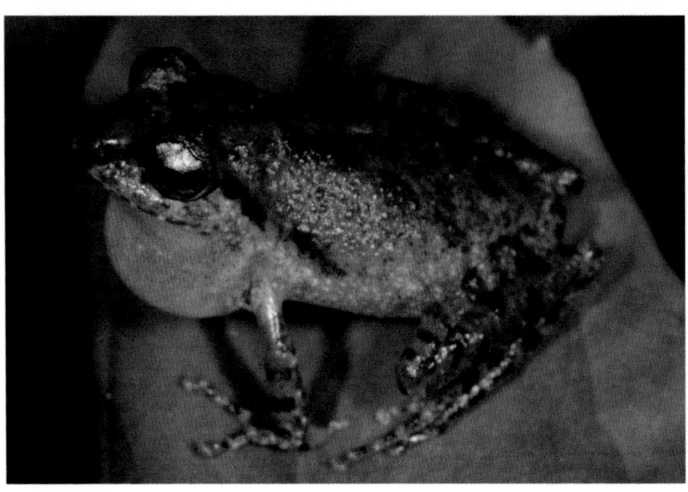

Fig. 1 Dorsolateral aspect

Fig. 2 *Lateral aspect*

Fig. 3 *Ventral aspect*

CHEEKY SHRUB FROG *Pseudophilautus procax*

(Kamule pellmethi pañduru mädiyã)

First described as *Philautus procax* by Kelum Manamendra-Arachchi and Rohan Pethiyagoda in 2005. Sri Lankan and Indian *Philautus* species were later placed in the genus *Pseudophilautus*, and thus its current name is *Pseudophilautus procax*.

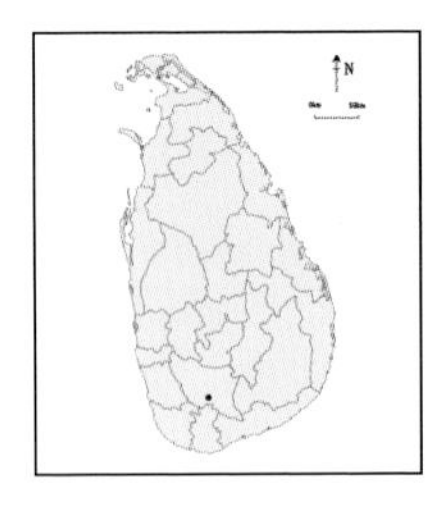

Size SVL 25–37mm.

Identification Features Small frog with slender body, dorsally concave head, pointed or oval snout in lateral aspect, rounded canthal edges, flat interorbital space, distinct tympanum and supratympanic fold, lateral dermal fringes on fingers and medially webbed toes. Both loreal region and internarial space concave. Tympanum oval, and vertically arranged. Lingual papilla present. Snout, interorbital area and sides of head smooth. Dorsum has horn-like spinules. Upper flanks smooth and lower flanks granular. Dorsal parts of forelimbs, thighs, shanks and feet smooth. Fine granules on throat and chest. Belly and undersides of thighs granular. Resembles *P. abundus*, but differs by medially webbed toes (vs fully webbed toes in *P. abundus*). (Figs 1 & 2).

Colour Dorsum pale brown with dark brown markings. Yellowish or pale mid-dorsal stripe may be present from tip of snout to vent in some individuals. Loreal and temporal areas dark brown. Yellow infraorbital patch. Colours of dorsum and venter meet in well-defined zone on flanks. Limbs and thigh dorsally dark brown. (Figs 1 & 2).

Habits Nocturnal species. Males can be seen vocalizing perched on leaves about 1–2m above ground level.

Habitat and Distribution Habitat generalist that can be seen in forest edges, and both open and closed canopy forests. Found near streams and marshy areas in forests and cardamom plantations. Restricted to elevations of 1,000m above sea level in the Rakwana mountain range.

Status Endemic.

IUCN Red List Category Least Concern.

Fig. 1 Dorsolateral aspect

Fig. 2 Lateral aspect, with infraorbital patch indicated by arrow

Puran Appu's Shrub Frog *Pseudophilautus puranappu*

(Puran Appuge pañduru mädiyã)

First described as *Pseudophilautus puranappu* by L. J. M. Wickramasinghe, D. R. Vidanapathirana, M. D. G. Rajeev, S. C. Ariyarathne, A. W. A. Chanaka, L. L. D. Priyantha, I. N. Bandara and N. Wickramasinghe in 2013.

Size SVL 25–37mm.

Identification Features Small to medium-sized frog with moderately elongated body, dorsally concave head, laterally and ventrally rounded, dorsally truncated snout, rounded canthal edge, flat interorbital space, distinct tympanum and supratympanic fold. Loreal region and internarial space concave. Tympanum oval, and vertically arranged. Lingual papilla present. Finger-tip discs have circummarginal grooves. Supernumerary tubercles on all fingers. Dorsal snout smooth with blunt tubercle and lateral snout smooth. Small tubercles on heels. Anterior dorsum weakly shagreened with few horny spinules, and posterior dorsum with few horny spinules and prominent tubercles. Tarsi, forearms and upper parts of flanks weakly tubercular and lower parts of flanks granular. Thighs smooth and legs tuberculated. Throat, chest, upper arms, forearms, thighs and tarsi weakly granular, and belly granular. Resembles *P. dimbullae*, from which it can be distinguished by dorsally concave head and lack of supernumerary tubercles on feet (vs dorsally convex head and presence of supernumerary tubercles on feet in *P. dimbullae*). (Figs 1 & 2).

Colour Dorsum uniform light brown with hourglass-shaped, maroon longitudinal band covering snout to vent. Faint stripe between eyes distinguishing fairly lighter triangular region on snout. Limbs dark brown with green tinge. Forelimbs, hindlimbs, fingers and toes have blackish maroon bars. (Figs 1 & 2). Ventral aspect off white marbled with dark brown.

Habits Active both by day and night. Commonly seen on rocks in shady places by day, and perching on 1m-high bushes at night.

Habitat and Distribution Common in places where there is no canopy cover in disturbed areas. Reported from highest elevations of 1,600–2,100m above sea level.

Status Endemic.

IUCN Red List Category Endangered.

Fig. 1 *Dorsolateral aspect*

Fig. 2 *Lateral aspect*

POLONNARUWA SHRUB FROG *Pseudophilautus regius*

(Rajarata pañduru mädiyã)

First described as *Philautus regius* by Kelum Manamendra-Arachchi and Rohan Pethiyagoda in 2005. Sri Lankan and Indian *Philautus* species were later placed in the genus *Pseudophilautus*, and thus the current name is *Pseudophilautus regius*.

Size SVL 18–27mm.

Identification Features Small frog with dorsally flat head, oval snout in lateral aspect, rounded canthal edges, concave loreal region, medially webbed toes, and distinct tympanum and supratympanic fold. Both interorbital and internarial spaces concave. Tympanum oval shaped and oblique. Supernumarary tubercles on palms and feet. Snout, interorbital area, sides of head, and anterior and posterior dorsum have horn-like spinules. Dorsum and upper flanks have horn-like spinules. Upper and lower flanks glandular. Dorsal parts of forelimbs, thighs and shanks have glandular warts. Feet dorsally smooth. Throat, chest, belly and undersides of thighs granular. Resembles *P. popularis* and can be distinguished by horny spinules on snout, sides of head and interorbital area (vs glandular warts on snout and sides of head and smooth interorbital area in *P. popularis*). (Figs 1 & 2).

Colour Dorsum highly variable and about six colour variations can be seen. Common colour is mixture of light chocolate-brown. Mid-dorsum dark brown with laterally dark brown head. Tympanum golden-yellow. Dorsal sides of thighs darker and contain brown cross-bars (Figs 1 & 2).

Habits Usually perches on leaves about 1m above ground level at night. Becomes active during the north-east monsoons (December–March), and female, while in amplexus, digs a small hole in humus, lays her eggs, then covers them up.

Habitat and Distribution Known from a variety of habitats, including forests, scrub and small thickets near paddy fields, vegetable plots and plant nurseries. One of two species of the genus *Pseudophilautus* known from dry zone of Sri Lanka. Reported from dry, shrubby habitats 20–600m above sea level, near rice fields around Anuradhapura, Polonnaruwa, Mihintale, Giritale, Wilpaththu National Park, Galnewa and Nilgala.

Status Endemic.

IUCN Red List Category Least Concern.

Fig. 1 Dorsolateral aspect

Fig. 2 *Lateral aspect*

Fig. 3 *Ventral aspect*

Reticulated Thigh Shrub Frog *Pseudophilautus reticulatus*

(Jalabha pañduru mädiyã)

First described as *Polypedates reticulatus* by Albert Günther in 1864. After several changes to the generic name (that is, to *Rhacophorus reticulatus*, *Philautus reticulatus* and *Kirtixalus reticulatus*), it is currently known as *Pseudophilautus reticulatus*.

Size 42–61mm.

Identification Features Medium-sized, elongated frog with dorsally convex or flat head, truncated or oval snout in lateral aspect, sharp canthal edges, concave loreal region, distinct tympanum and supratympanic fold, two short spines (that is, tarsal tubercle and calcar) on heels, fingers with dermal fringes, fully webbed toes, and reticulated pattern on posteriors of thighs. Interorbital and internarial spaces convex or concave. Tympanum oval shaped. Rounded lingual papilla. Dorsal surface finely tuberculated. Chin smooth and chest granular. Ventral aspects of upper arms, abdomen, and ventral and posterior thighs rough and granular. Dorsum, snout, interorbital space, sides of head, and upper parts of flanks shagreened with a few scattered glandular warts. Lower flanks granular. Dorsal parts of forelimbs shagreened. Thighs, shanks and feet smooth. Throat and chest smooth or granular. Belly and undersides of thighs granular. Resembles *P. maia* and *P. papillosus*. Can be distinguished from *P. maia* by absence of calcars on heels in *P. maia*. Distinguished from *P. reticulatus* by smaller lingual papilla (vs larger lingual papilla in *P. papillosus*). (Figs 1 & 2).

Colour Dorsum and dorsal sides of head brown with indistinct dark brown patches. Limbs brown with indistinct dark brown cross-bars. Canthal edges, edges of upper eyelids, and supratympanic fold orange-brown. Belly ashy-brown with dark brown patches. Flanks ashy-brown marbled with dark brown (Figs 1 & 2). Thighs marbled (reticulated) with dark brown (Fig. 3). Lateral fringes on fingers and toes pale orange.

Habits Common habitat generalist that is mostly active at night. Can be seen on the ground by day and at night, on trees at a height of 10–20m above the ground. Calls can be heard from the tree-tops, mostly at night.

Habitat and Distribution Restricted to lowland rainforests in south-west wet zone of Sri Lanka, including home gardens and plantations at elevations of 30–900m. Known from Kosmulla, Dediyagala, Yagirala, Induruwa, Norton Bridge, Haycock, Kudawa, Kanneliya, Ambagamuwa, Udamaliboda and Hapugasthenna.

Status Endemic.

IUCN Red List Category Endangered.

Farnland Shrub Frog *Pseudophilautus rugatus*

(Fernland panduru mädiya)

First described as *Rhacophorus rugatus* by Ernst Ahl in 1927. After several changes to the generic name (including to *Philautus rugatus*), currently known as *Pseudophilautus rugatus*. Known only from the holotype specimen deposited in the Zoological Museum of Berlin, bearing the museum number ZMB8557. There have been no sightings of the species since it was described in 1927, and it is now considered to be extinct.

Status Endemic.

IUCN Red List Category Extinct.

Fig. 1 Dorsolateral aspect

Fig. 2 Lateral aspect, with short spines indicated by arrow

Fig. 3 Reticulation on thigh indicated by arrow

KANDIAN SHRUB FROG *Pseudophilautus rus*
(Nuwara pañduru mädiyã)

First described as *Philautus rus* by Kelum Manamendra-Arachchi and Rohan Pethiyagoda in 2005. Sri Lankan and Indian *Philautus* species were later placed in the genus *Pseudophilautus*, and hence the current name is *Pseudophilautus rus*.

Size SVL 20–24mm.

Identification Features Small frog with distinct tympanum and supratympanic fold, dorsally flat head, oval snout in lateral aspect, rounded canthal edges, concave loreal region, glandular dorsum with prominent warts, and medially webbed feet. Both interorbital and internarial spaces flat. Tympanum oval, and oblique. Throat, chest, belly, flanks and undersides of thighs granular. Resembles *P. conniffae* and *P. silvaticus*. Differs from *P. silvaticus* by 'V'-shaped tubercular pattern on dorsum (vs absence of 'V'-shaped tubercular pattern on dorsum in *P. silvaticus*). Differs from *P. conniffae* by smooth interorbital area (vs interorbital area with isolated glandular warts in *P. conniffae*). (Figs 1 & 2).

Colour Dorsum dark to light brown with darker patches. Upper flanks chestnut-brown with dark brown patches, and lower flank pale yellow with dark brown patches. Dorsal and lateral parts of forelimbs, and dorsal parts of thighs, shanks and feet brown with dark brown cross-bars. (Figs 1 & 2). Anterior and posterior surfaces of thighs and flanks black with white spots (Fig. 3). Thighs brown with dark brown mottling on posterior edges (Fig. 3).

Habits Hides by day under leaf litter and inside crevices in trees and cracks in the soil, emerging at dusk. Very active at dusk and common during the rains. Males call from leaves and branches less than 1m above ground level. Common in anthropogenic habitats and occasionally may be encountered inside houses at night.

Habitat and Distribution Common habitat generalist that occupies undisturbed forests, secondary forests, plantations and densely wooded home gardens. Known from Kandy, Peradeniya, Gelioya, Pilimathalawa, Loolkandura, Gampola, Nawalapitiya, Ambagamuwa, Kithulgala, Pussellawa and Ramboda at elevations of 200–1,200m above sea level.

Status Endemic.

IUCN Red List Category Near Threatened.

Fig. 1 Dorsolateral aspect

Fig. 2 Dorsolateral aspect

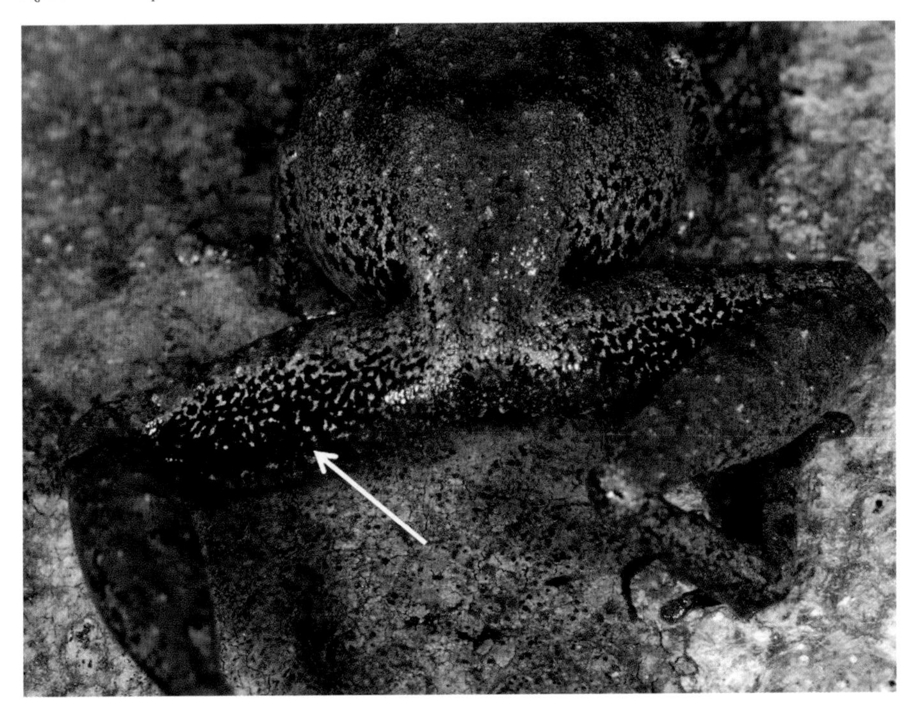

Fig. 3 Posterior side and thighs, with dark brown mottling indicated by arrow

Samarakoon's Shrub Frog *Pseudophilautus samarakoon*

(Samarakoonge pañduru mädiyā)

First described as *Pseudophilautus samarakoon* by L. J. M. Wickramasinghe, D. R. Vidanapathirana, M. D. G. Rajeev, S. C. Ariyarathne, A. W. A. Chanaka, L. L. D. Priyantha, I. N. Bandara and N. Wickramasinghe in 2013.

Size SVL 22–25mm.

Identification Features Small frog with broad body, dorsally convex head, laterally truncated and dorsally mucronate snout, rounded canthal edges, distinct tympanum and supratympanic fold, concave loreal region, convex interorbital region and medially webbed feet. Internasal space flat. Tympanum elliptical. Fingertip discs have distinct basal and circummarginal grooves. Dermal fringes on all toes and on insides of all fingers. Fingers and toes have supernumerary tubercles. Blunt calcars and small tubercles on heels. Dorsum, upper parts of flanks, forearms and dorsal sides of legs have tubercules. Skin between eyes tubercular, with prominent tubercle in centre. Median dermal ridge from tip of snout to vent, and prominent tubercle on centre of snout. Pair of prominent tubercles above nostrils, and tubercles prominent on upper eyelids. Lower parts of flanks granular. Upper arms, hands, and inner and outer thighs smooth. Throat, chest, thighs, legs and tarsi weakly granular, and belly granular at ventral aspect. Resembles *P. silus*, but can be differentiated by rounded canthal edges and truncated snout in lateral view (vs sharp canthal edges and oval snout in lateral view in *P. silus*). (Figs 1 & 2).

Colour Dorsum uniform creamy light brown with dark brown patches. Dark brown band between eyes and another in front. Large, dark brown, 'W'-shaped patch on mid-dorsum, with creamy light brown, oval centre in anterior region and two symmetric dark brown diagonal lines in posterior region. Body cream in both ventral and lateral aspects, and limbs dorsally creamy light brown. Forelimbs, hindlimbs, fingers and toes have brown cross-bands (Figs 1 & 2).

Habits Arboreal species that can be seen on the forest floor by day. Ascends to low vegetation close to streams about 2m from the ground.

Habitat and Distribution Found at elevations of 1,000–1,400m in submontane rainforests in the Sri Pada (Peak Wilderness) Sanctuary.

Status Endemic.

IUCN Red List Category Critically Endangered.

Fig. 1 Dorsolateral aspect

Fig. 2 Dorsolateral aspect

Muller's Shrub Frog *Pseudophilautus sarasinorum*

(Mularge pañduru mädiyã)

First described as *Ixalus sarasinorum* by Friedrich Muller in 1887. After several changes to the generic name (that is, to *Micrixalus sarasinorum*, *Staurois sarasinorum* and *Philautus sarasinorum*), it is currently known as *Pseudophilautus sarasinorum*.

Size SVL 26–40mm.

Identification Features Small to medium-sized frog with dorsally flat head, rounded snout in lateral aspect, rounded canthal edges, distinct tympanum and supratympanic fold, flat interorbital space and fully webbed toes. Loreal region and internarial spaces concave. Tympanum oval shaped. Lingual papilla present. Snout, interorbital area, sides of head and dorsum have glandular warts. Upper flanks contain glandular folds. Lower flanks granular in female, and horn-like spinules scattered on dorsum of male. Dorsal parts of forelimbs, thighs and shanks smooth. Feet and throat have glandular warts. Chest and undersides of thighs smooth. Belly granular. Resembles *P. procax*, *P. macropus* and *P. abundus*, from which it can be differentiated by rounded snout in lateral aspect, and absence of lateral dermal fringes on fingers (vs by pointed or oval snout in lateral aspect, and lateral dermal fringes on fingers in *P. procax* and *P. abundus*). (Figs 1 & 2).

Colour Dorsum and dorsal head light to dark brown. Dark interorbital bar and dark brown, 'W'-shaped symmetrical marking on mid-dorsum. Limbs dorsally brown with dark cross-bars. Yellow spot below eye. (Figs 1 & 2). Abdomen yellow.

Habits Nocturnal species found on low (<1.5m) branches overhanging water, mainly at night. By day hides in rock crevices on margins of streams.

Habitat and Distribution Occurs on boulders or branches close to water bodies in both closed and open-canopy forests, and anthropogenic habitats. Known from the Central Hills and Knuckles Mountain Range, at elevations of 800–1,300m, including at Hanthana, Hunasgiriya, Hapugasthenna, Thangappuwa, Agra-Bopath, Bogowantalawa and Corbet's Gap.

Status Endemic.

IUCN Red List Category Endangered.

Fig. 1 Dorsolateral aspect

Fig. 2 Lateral aspect

SCHMARDA'S SHRUB FROG *Pseudophilautus schmarda*

(Gorahadi pañduru mädiyã)

First described by Edward Ferderic Kelaart in 1854 as *Polypedates schmarda*. The first amphibian species described by a Sri Lankan. After several generic name changes (that is, to *Ixalus poecilopleurus, I. schmardae, I. schmardanus, Philautus schmardanus, Rhacophorus schmardanus, Theloderma schmardanus, T. schmardanum, T, schmarda, Philautus schmarda* and *Kirtixalus schmarda*), it is currently known as *Pseudophilautus schmarda*.

Size SVL 17–30mm.

Identification Features Small frog with stout body, dorsally convex head, obtusely pointed snout in lateral view, rounded canthal edges, distinct tympanum and supratympanic fold, tuberculated tarsal fold, calcars at heels and medially webbed toes. Interorbital space convex. Loreal region and internarial space concave. Tympanum oval shaped. Fingers have lateral dermal fringes. Snout contains glandular warts. Glandular folds, glandular warts and horn-like spinules in interorbital area, and on dorsum and upper flanks. Sides of head have glandular warts. Lower flanks have glandular warts and horn-like spinules. Throat and dorsal parts of forelimbs, thighs, shanks and feet have glandular warts. Chest, belly and undersides of thighs granular. Resembles *P. dilmah* and *P. hankeni*. Differs from *P. dilmah* by obtusely pointed snout and sharp canthal ridge (vs rounded snout when viewed laterally and rounded canthal ridge in *P. dilmah*). Differs from *P. hankeni* by absence of dermal fringes on fingers (vs dermal fringes on fingers in *P. hankeni*). (Figs 1 & 2).

Colour Dorsum highly variable, from dark green to light green and red-brown with dark irregular patches. Dorsal aspects of limbs have patches. (Figs 1 & 2).

Habits Nocturnal species that perches on low vegetation (about 1–2m above the ground), on moss, and lichen-covered boulders and roots. Hides underneath leaf litter, logs and stones, and inside crevices by day. Begins its characteristic call at dusk (about 5.30 p.m.), and occasionally calls even during the day. Cryptic colour and skin texture camouflage it well when sitting on lichen-covered boulders and tree trunks.

Habitat and Distribution Mainly confined to submontane and montane forests, and occasionally found in anthropogenic habitats. Generally reported from the Central Hills at elevations of 810–2,300m. One of the most common amphibians inhabiting Horton Plains. Also known from the Peak Wilderness, Nuwara Eliya, Hakgala, Haputhale and Agrabopath.

Status Endemic.

IUCN Red List Category Category Endangered.

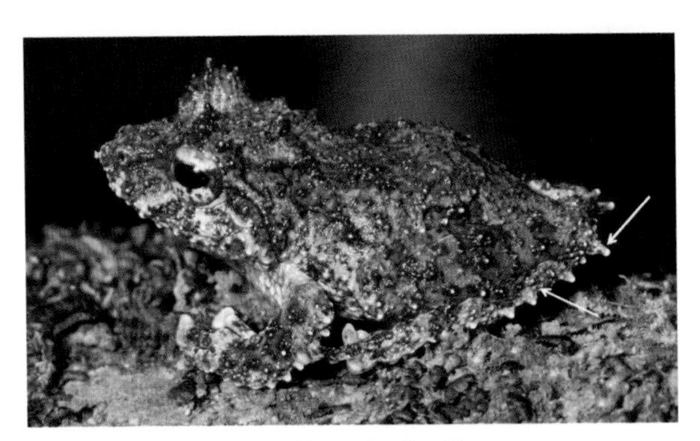

Fig. 1 Dorsolateral aspect, with horn-like spinules indicated by arrows

Fig. 2 Lateral aspect, with horn-like spicules indicated by arrow

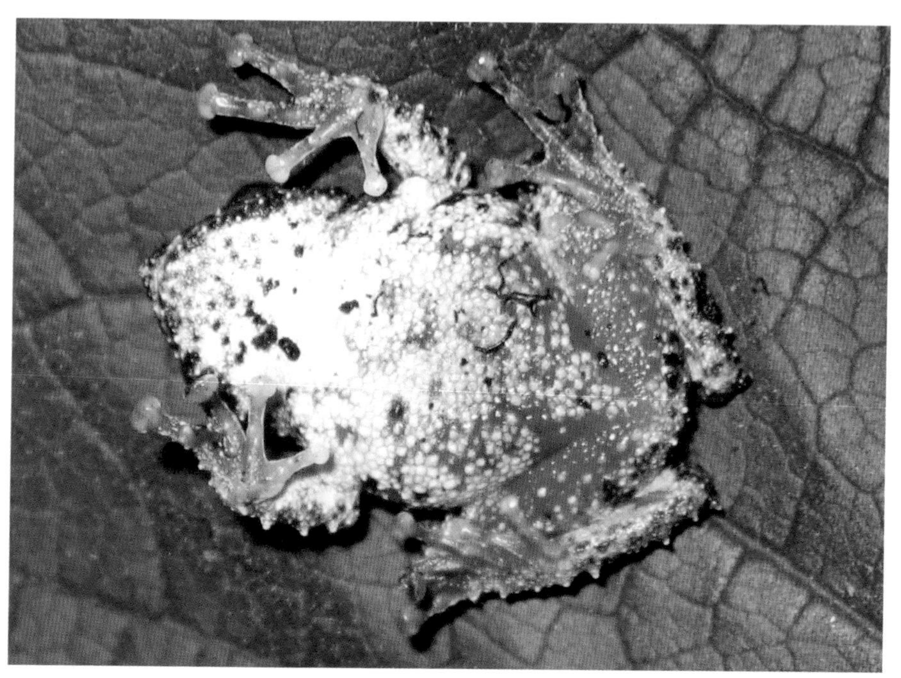

Fig. 3 Ventral aspect

Schneider's Shrub Frog *Pseudophilautus schneideri*

(Schneiderge panduru mediya)

First described as *Pseudophilautus schneideri* by Madhava Meegaskumbura and Kelum Manamendra-Arachchi in 2011.

Size 19–30mm.

Identification Features Small frog with somewhat flattened, elongated body, laterally convex head, obtusely pointed snout in dorsal view and pointed snout in lateral view, sharp canthal edges, distinct tympanum and supratympanic fold, and medially webbed toes. Interorbital space convex. Loreal region and internasal space flat. Tympanum oval shaped. Fingers lack lateral dermal fringes. Supernumerary tubercles on toes and foot. Small tubercles with horny spinules on dorsal and lateral aspects of head and body. Lower flanks granular. Dorsolateral fold absent. Dorsal and lateral parts of upper arms, lower arms, thighs, shanks and feet smooth. Throat, chest and belly granular. Resembles *P. folicola* and distinguished from it by obtusely pointed snout in lateral aspect, flat loreal region, ventrally smooth thighs and absence of lateral dermal fringes on fingers and supernumerary tubercles on palms (vs rounded or truncated snout in lateral aspect, concave loreal region, ventrally granular thighs, lateral dermal fringes on fingers and supernumerary tubercles on palms in *P. folicola*. (Figs 1–3).

Colour Dorsum beige to light brown with minute black and dark brown spots. Head laterally dark brown. Limbs and flanks pale brown with dark brown spots. Limbs have dark brown cross-bars. (Figs 1–3).

Habits Found perched on leaves of shrubs 0.5–1.5m above the ground in open and partially open habitats. By day can be seen among leaf litter.

Habitat and Distribution Commonly found in forests, forest edges and anthropogenic habitats such as well-shaded home gardens

Fig. 1 Dorsolateral aspect

and tea plantations. Widely distributed in wet zone at elevations of 30–700m above sea level. Reported from Ampitiya-Kandy, Gannoruwa, Gampola, Nawalapitiya, Ambagamuwa, Kithulgala, Sinharaja, Kanneliya, Elpitiya, Pituwela, Udawattakele, Udamaliboda and Kudawa areas.

Status Endemic.

IUCN Red List Category Vulnerable.

Fig. 2 Dorsolateral aspect

Fig. 3 Lateral aspect

Annandale's Shrub Frog/Tiny Red Shrub Frog

Pseudophilautus semiruber

(Annandalge panduru mädiyã)

First described as *Ixalus semiruber* by Thomas Nelson Annandale in 1913. After several changes to the generic name (that is, to *Rhacophorus semiruber, Ixalus semiruber* and *Philautus semiruber*), it is currently recognized as *Pseudophilautus semiruber*.

Size SVL 12–18mm.

Identification Features Very small frog with stout body, laterally convex head, oval snout in dorsal and lateral views, rounded canthal edges, oval-shaped tympanum, distinct supratympanic fold and medially webbed toes. Loreal region and internarial space flat. Interorbital space concave. Fingers without lateral dermal fringes. Supernumerary tubercles on fingers, palms and toes. Dorsal and lateral parts of head, body, upper arms, lower arms, thighs, shanks and feet smooth. Narrow dermal ridge extends from mid-dorsum to back of head, then to vent. Throat, chest, belly and undersides of thighs smooth. Resembles *P. simba* and differentiated from it by feebly distinct supratympanic fold, medially webbed toes, and smooth throat, chest and belly (vs distinct supratympanic fold, rudimentarily webbed toes and granular throat, chest and belly in *P. simba*). (Figs 1 & 2).

Colour Dorsum grey-brown. Head laterally dark brown. Interorbital region grey. Whole upper arm and proximal half of lower arm dorsally red with black band extending anterodorsally from base of upper arm to proximal half of lower arm. Flanks ashy-brown with a few white patches outlined in red. (Figs 1 & 2). Chest and belly reddish-brown with white patches (Fig. 3).

Habits Assumed to be a diurnal species that inhabits the forest floor.

Habitat and Distribution Restricted to montane forests at 1,600m above sea level. Originally described from a single specimen

Fig. 1 Dorsolateral aspect

collected at Pattipola (about 1,850m above sea level), but was rediscovered in Agra-Bopath forest. Also known from Rilagala Forest Reserve.

Status Endemic.

IUCN Red List Category Endangered.

Fig. 2 Dorsolateral aspect

Fig. 3 Ventral aspect

PUG-NOSED SHRUB FROG *Pseudophilautus silus*

(Mukkang hombu panduru mädiyã)

First described as *Philautus silus* by Kelum Manamendra-Arachchi and Rohan Pethiyagoda in 2005. Sri Lankan and Indian *Philautus* species were later placed in the genus *Pseudophilautus*, and thus the current name is *Pseudophilautus silus*.

Size 35–52mm.

Identification Features Medium-sized frog with stout body, dorsally convex head, oval snout in lateral aspect, sharp canthal edges, concave loreal region, oval-shaped tympanum, distinct supratympanic fold and medially webbed toes. Interorbital and internarial spaces flat. Lateral dermal fringes on fingers. Dorsal and lateral parts of snout and interorbital area smooth. Sides of head, dorsum and upper flanks contain glandular warts. Dorsal parts of forelimbs, thighs, shanks and feet smooth. Throat, chest, belly, undersides of thighs and lower flanks granular. Resembles *P. fulvus* and can be distinguished by oval snout in lateral aspect (vs rounded snout in lateral aspect in *P. fulvus*). (Figs 1–3).

Colour Dorsum pale ashy-brown to greyish-green, with two dark brown dorsolateral bands. Head laterally darker. Dark brown interorbital bar. Flanks yellow with dark brown patches. Limbs have pale brown dorsal cross-bars. (Figs 1–3).

Habits Nocturnal species, with adults found at night close to stream banks, perched on branches about 1–4m above ground level.

Habitat and Distribution Habitat generalist living in closed canopy rainforests and

Fig. 1 Dorsolateral aspect

anthropogenic habitats such as tea estates. Known from lower Central Hills at elevations of 1,000–1,600m. Known from Agarapatana, Haputale, on either side of the Horton Plains plateau, Loolecondera and Mandaram Nuwara.

Status Endemic.

IUCN Red List Category Endangered.

Fig. 2 Dorsolateral aspect

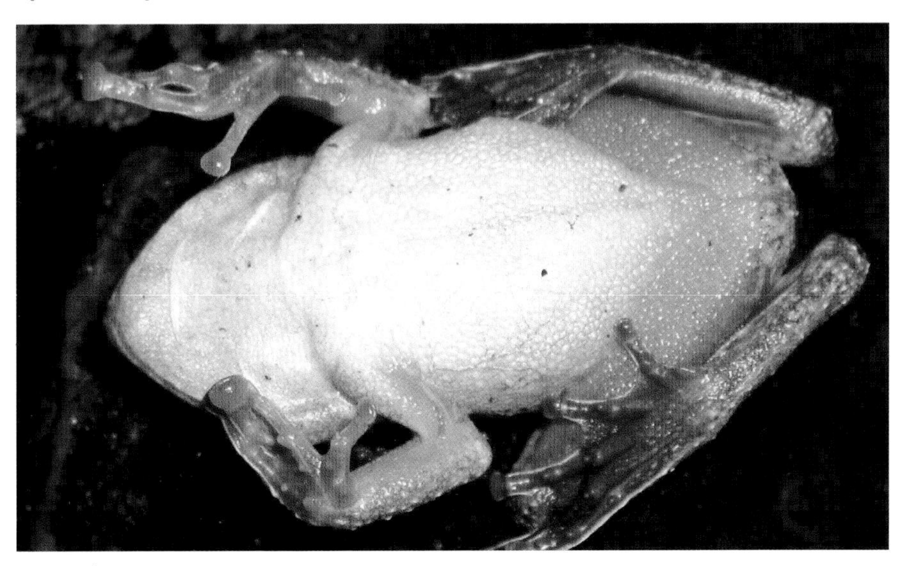

Fig. 3 Ventral aspect

FOREST SHRUB FROG *Pseudophilautus silvaticus*

(Mukalan pañduru mädiyã)

First described as *Philautus silvaticus* by Kelum Manamendra-Arachchi and Rohan Pethiyagoda in 2005. Following the placement of all Sri Lankan and Indian *Philautus* in the genus *Pseudophilautus*, its current name is *Pseudophilautus silvaticus*.

Size SVL 24–31mm.

Identification Features Small frog with elongated body, dorsally convex head, oval-shaped snout in lateral aspect, sharp canthal edges, concave loreal region, rounded tympanum, distinct supratympanic fold and medially webbed toes. Interorbital and internarial space flat. Lingual papilla on tongue. Snout, interorbital area, sides of head, anterior dorsum, dorsal parts of forelimbs, upper flanks, shanks and feet contain glandular warts. Posterior dorsum smooth. Lower flanks granular. Thighs dorsally smooth. Throat, chest, belly and undersides of thighs granular in male, and throat and chest smooth in female. Resembles *P. conniffae* and *P. rus*. Differs from *P. conniffae* by skin on head being co-ossified with cranium and by absence of 'V'-shaped tubercular pattern on dorsum (vs skin on head not co-ossified with cranium and presence of 'V'-shaped tubercular pattern on dorsum in *P. conniffae*). Differs from *P. rus* by lingual papilla and absence of 'V'-shaped tubercular pattern on dorsum (vs absence of lingual papilla and presence of 'V'-shaped tubercular pattern on dorsum in *P. rus*). (Figs 1 & 2).

Colour Dorsum grey-brown to dark brown with several black markings. Canthal edges orangish-brown. Loreal and temporal area brownish-grey with black markings. Flanks grey-brown with a few black spots. Limbs grey-brown with dark brown cross-bars. Thighs black with dark brown and white spots. (Figs 1 & 2).

Habits Nocturnal species in which adult males can be seen calling at night, perched on low vegetation about 30cm–1m above ground level. Seen among leaf litter by day.

Habitat and Distribution Habitat specialist that dwells in closed canopy rainforests and cardamom plantations. Restricted to heights of 510–1,240m in the Rakwana mountain range. Known from Morningside, Enasalwatta and Lankagama in the Sinharaja World Heritage Site.

Status Endemic.

IUCN Red List Category Endangered.

Fig. 1 Dorsolateral aspect

Fig. 2 Lateral aspect

Sinharaja Shrub Frog *Pseudophilautus simba*

(Sinharaja panduru mädiyã)

First described as *Philautus simba* by Kelum Manamendra-Arachchi and Rohan Pethiyagoda in 2005. All Sri Lankan and Indian *Philautus* species were subsequently placed in the genus *Pseudophilautus*, and thus the current name is *Pseudophilautus simba*.

Size SVL 12–16mm.

Identification Features Very small frog with short body, dorsally convex head, truncated snout in lateral aspect, rounded canthal edges, convex interorbital space, oval-shaped tympanum, distinct supratympanic fold and rudimentarily webbed toes. Loreal region and internarial space flat. Lateral dermal fringes absent on fingers. Snout, interorbital area and posterior dorsum smooth. Dorsum and lateral side of head has glandular warts. Dorsal parts of forelimbs, thighs, shanks and feet smooth. Throat and chest granular. Belly, undersides of thighs and lower flanks granular. Resembles *P. semiruber*, but can be distinguished by distinct supratympanic fold, rudimentarily webbed toes and granular throat, chest and belly (vs feebly distinct supratympanic fold, medially webbed toes and smooth throat, chest and belly in *P. semiruber*). (Figs 1 & 2).

Colour Dorsum and dorsal aspect of head light brown to chestnut-brown. Inguinal zone ashy-brown. Flanks dark brown with white or light blue patches. Thighs and shanks dorsally chestnut-brown with dark brown cross-bars. Feet light brown dorsally and dark brown ventrally. (Figs 1 & 2). Chest and belly brown mottled with white spots (Fig. 3).

Habits Ground-dwelling leaf-litter species. Seen slowly crawling along forest floor by day.

Habitat and Distribution Habitat specialist found in closed canopy habitats that include submontane forests and cardamom plantations located in submontane forests. Restricted to elevations above 1,000m in the Rakwana Mountains. Known from Morningside area and Enasalwatta in the Sinharaja forest.

Status Endemic.

IUCN Red List Category Critically Endangered.

Fig. 1 Dorsolateral aspect

Fig. 2 Dorsolateral aspect

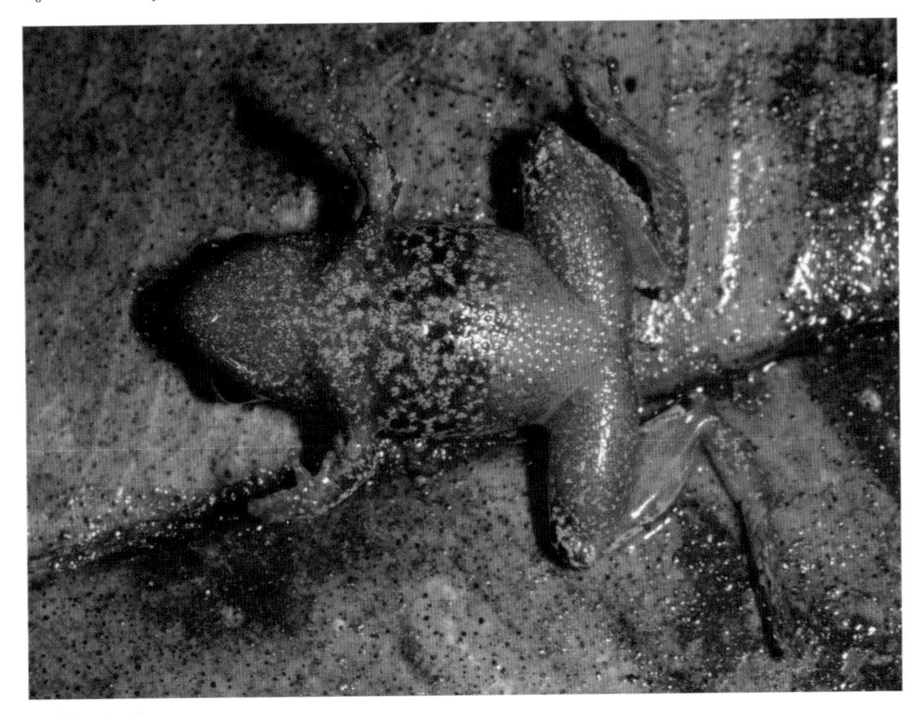

Fig. 3 Dorsolateral aspect

SRI LANKA SHORT-HORNED SHRUB FROG *Pseudophilautus singu*
(Singu pañduru mädiyã)

First described as *Philautus singu* by Madhava Meegaskumbura, Kelum Manamendra-Aarachchi and Rohan Pethiyagoda in 2009. Following the placement of Sri Lankan and Indian *Philautus* species in the genus *Pseudophilautus*, its current name is *Pseudophilautus singu*.

Size SVL 16–30mm.

Identification Features Small frog with short body, laterally convex head, obtusely pointed snout in dorsal view, rounded snout in lateral view, rounded canthal edges, convex interorbital space, oval-shaped tympanum, distinct supratympanic fold and medially webbed toes. Loreal region and internasal space concave. Lingual papilla present. Supernumerary tubercles on fingers. Dorsal and lateral regions of snout, between eyes, sides of head, anterior part of back, posterior part of back, both upper and lower flanks, upper arms, lower arms, thighs, shanks and feet have scattered glandular and horny tubercles. Throat, chest, belly and ventral sides of thighs granular. Resembles *P. decoris* and *P. mittermeieri*. Differs from them by prominent tubercle on upper eyelid, (vs absence in *P. mittermeieri* and *P. decoris*), rounded snout in lateral aspect (vs pointed snout in *P. mittermeieri* and obtusely pointed snout in *P. decoris*), absence of lateral dermal fringes on fingers (vs presence in *P. mittermeieri* and *P. decoris*), and absence of tarsal fold (vs presence in *P. mittermeieri* and *P. decoris*). (Figs 1 & 2).

Colour Dorsum variable, from olive-brown to greenish-brown. 'W'-shaped dark brown marking on mid-back. Upper flanks brown and lower flanks yellow with brown pigments. Dorsal and lateral parts of forelimbs pale yellow with brown pigments and without distinct cross-bars. Thighs and shanks pale brown with three dark brown cross-bars. (Figs 1 & 2).

Habits Can be seen perched on leaves of shrubs about 0.5–1.5m above ground level. By day found among leaf litter on the forest floor.

Habitat and Distribution Occurs in lowland rainforests and adjoining vegetation at 60–700m above sea level. Known from Kottawa Forest Reserve, Sinharaja Forest Reserve, Kanneliya-Dediyagala-Nakiyadeniya forests, Udamaliboda and Kitulgala.

Status Endemic.

IUCN Red List Category Endangered.

Fig. 1 Dorsolateral aspect

Fig. 2 Dorsal aspect

Siril Wijesundara's Shrub Frog *Pseudophilautus sirilwijesundarai*

(Siril Wijesundarage pañduru mädiyã)

First described as *Pseudophilautus sirilwijesundarai* by L. J. M. Wickramasinghe, D. R. Vidanapathirana, M. D. G. Rajeev, S. C. Ariyarathne, A. W. A. Chanaka, L. L. D. Priyantha, I. N. Bandara and N. Wickramasinghe in 2013.

Size SVL 32–33mm.

Identification Features Small frog with elongated body, dorsally convex head, truncated snout in dorsal view, rounded snout in lateral and ventral views, rounded canthal edges, concave loreal region, slightly distinct tympanum, distinct supratympanic fold and medially webbed feet. Internarial space flat and interorbital space convex. Tympanum oval shaped. Tips of digits have distinct basal and circummarginal grooves. Dermal fringes on all fingers. Supernumerary tubercles prominent on palms, feet, toes and fingers. Dorsum and snout shagreened and head laterally weakly shagreened, with a few tubercles near gape of mouth. Interorbital area has prominent tubercle. Median dermal ridge from tip of snout to back of head. Inner, dorsal and outer surfaces of thighs smooth. Legs, tarsi and feet smooth. Throat shagreened, chest weakly granular and belly coarsely granular. Upper arms and forearms weakly granular and thighs granular ventrally. (Figs 1 & 2).

Colour Dorsum dark brown with green tinge and maroon blotches placed symmetrically. Maroon patch below eye on lateral sides of head. Snout and interorbital area maroon with well-defined triangle. Limbs dorsally olive-green, and forelimbs, hindlimbs, fingers and toes have prominent maroon cross-bands. Throat, hands and feet dark brown and belly off white with brownish blotches. (Figs 1 & 2).

Habits Found on moss-covered branches in the canopy about 10m above the ground.

Habitat and Distribution Known from montane cloud forests in the Peak Wilderness Sancuary at 1,600–1,700m.

Status Endemic.

IUCN Red List Category Critically Endangered.

Fig. 1 Dorsolateral aspect

Fig. 2 Dorsal aspect

Grubby Shrub Frog *Pseudophilautus sordidus*
(Anduru lapawan pañduru mädiyã)

First described as *Philautus sordidus* by Kelum Manamendra-Arachchi and Rohan Pethiyagoda in 2005. All Sri Lankan and Indian *Philautus* species were subsequently placed in the genus *Pseudophilautus*. Thus the current name is *Pseudophilautus sordidus* (Manamendra-Arachchi & Pethiyagoda, 2005).

Size SVL 27–39mm.

Identification Features Small frog with elongated body, dorsally convex head, blunt snout in lateral aspect, rounded canthal edges, concave loreal region, convex interorbital space, flat internarial space, distinct oval-shaped tympanum, distinct supratympanic fold and medially webbed toes. Lateral dermal fringes absent on fingers. Snout, interorbital area, sides of head, dorsum and upper flanks have glandular warts. Lower flanks granular. Dorsal parts of forelimbs, thighs and shanks have glandular warts. Throat and chest smooth. Belly and ventral aspects of thighs granular. Nuptial pads absent in male. Resembles *P. macropus* but differs by presence of medially webbed toes (vs fully webbed toes in *P. macropus*) (Figs 1–3).

Colour Dorsal and lateral aspects of head and body vary from chestnut to dark or ashy-brown with dark brown symmetrical patches. Interorbital bar black or dark brown. Series of black or dark brown symmetrical markings on dorsum. Dorsal and lateral parts of forelimbs chestnut or brown with dark brown cross-bars. Thighs, shanks and feet chestnut-brown with dark brown cross-bars. (Figs 1–3).

Habits Nocturnal species that hides in rock crevices and fallen logs.

Habitat and Distribution Occurs around streams that traverse both closed and open canopy forests and anthropogenic habitats. Can also be seen inside houses. Reported at elevations of 80–1,060m including at Kanneliya, Haycock, Millawa, Morningside, Hanthana, Kudawa, Kithulgala, Welikanna, Labugama, Sinharaja and Yagirala.

Status Endemic.

IUCN Red List Category Not Evaluated.

Fig. 1 Dorsolateral aspect

Fig. 2 Dorsolateral aspect

Fig. 3 Ventral aspect

Steiner's Shrub Frog *Pseudophilautus steineri*

(Steinerge pañduru mädiyã)

First described as *Philautus steineri* by Kelum Manamendra-Arachchi and Rohan Pethiyagoda in 2005. Subsequently placed in the genus *Pseudophilautus*, and thus its current name is *Pseudophilautus steineri*.

Size SVL 27–39mm.

Identification Features Small to medium-sized frog with stout body, dorsally flat head, rounded snout in lateral aspect, sharp canthal edges, flat interorbital space, distinct tympanum and supratympanic fold, and medially webbed feet. Loreal region and internasal space concave. Lateral dermal fringes on fingers. Supernumerary tubercles on palms and pes. Snout, interorbital space, dorsum and upper flanks have glandular warts and horny spinules. Sides of head smooth or have glandular warts. Dorsal parts of forelimbs, thighs, shanks and pes smooth with scattered glandular warts. Throat, chest, belly and undersides of thighs granular. Feebly defined dermal fringe extends from tip of snout to posterior dorsum. Resembles *P. microtympanum*, from which it can be distinguished by laterally rounded snout and sharp canthal edges (vs laterally oval snout and rounded canthal edges in *P. microtympanum*). (Figs 1–3).

Colour Dorsum brown and dorsolateral area light greenish. Upper flanks light brown with black patches and lower flanks ashy-green. Dark brownish-black stripe on upper flanks. Interorbital bar dark brown. Limbs dorsally brown with dark brown cross-bars. Digits pale ashy-brown. (Figs 1–3).

Habits Arboreal species that can be seen on moss-covered rocks, logs and low vegetation.

Habitat and Distribution Occurs in submontane and montane forests, and cardamom plantations of the Knuckles Mountain Range at above 1,100m.

Status Endemic.

IUCN Red List Category Endangered.

Fig. 1 Dorsolateral aspect

Fig. 2 Dorsal aspect

Fig. 3 Dorsal aspect

Spotted Shrub Frog *Pseudophilautus stellatus*

(Pulli sahitha pañduru mädiyã)

First described as *Polypedates stellata* by Edward Frederick Kelaart in 1853. After several changes to the generic name (that is, to *Rhacophorus stellatus, Kirtixalus stellatus* and *Philautus stellatus*), currently known as *Pseudophilautus stellatus*. Known only from the lost holotype and original description until Mendis Wickramasinghe and his team rediscovered it at Sri Pada Sanctuary in 2009.

Size SVL 39–55mm.

Identification Features Medium-sized frog with rounded snout in lateral, dorsal and ventral aspects; indistinct tympanum; concave head from above; concave internarial space; rounded canthal edges, concave loreal region; concave interorbital space, third and fourth fingers with split (bifid) distal subarticular tubercles and medially webbed toes. Supratympanic fold absent. Dermal fringes on insides of all fingers. Skin on dorsal and lateral snout, head and entire dorsum weakly shagreened; upper flanks shagreened to weakly areolate; lower flanks weakly areolate to granular; upper arms, forearms and hands weakly shagreened; inner thighs dorsally and outer thighs weakly shagreened; legs, tarsi and feet weakly shagreened; single prominent large, blunt tubercle on each heel. Skin on ventral side of body: throat and chest weakly granular; belly granular; upper arms weakly granular, forearms granular; thighs granular, legs smooth, tarsi weakly granular. (Figs 1 & 2).

Colour Dorsum of body, legs and lower arms bright green with intermittent pinkish-white spots outlined by thin dark brown lines. Flanks have transverse dark brown bands on white. Chest and belly all pinkish-white. (Figs 1 & 2).

Habits Nocturnal, arboreal species that can be seen on vegetation at 1–10m from the ground.

Habitat and Distribution Known only from Nuwara Eliya and montane cloud forests of Sri Pada Sanctuary at 1,500–1,600m above sea level.

Status Endemic.

IUCN Red List Category Critically Endangered.

Fig. 1 Dorsolateral aspect

Fig. 2 Dorsolateral aspect

Orange Canthal Shrub Frog *Pseudophilautus stictomerus*

(Thembili hombu pañduru mädiyã)

First described as *Ixalus stictomerus* by Albert Günther in 1876. After several changes to the generic name (that is, to *Rhacophorus stictomerus* and *Philautus stictomerus*), currently known as *Pseudophilautus stictomerus.*

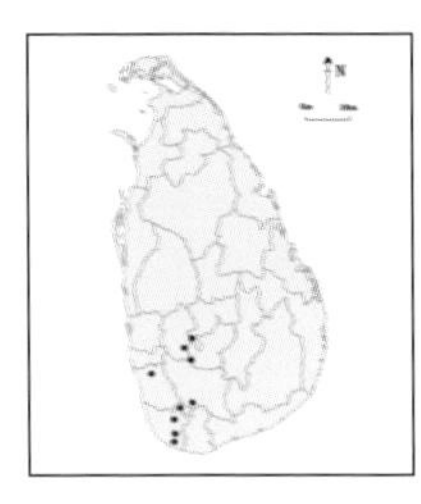

Size SVL 27–39mm.

Identification Features Small frog with elongated body, elongated, dorsally flat or convex head, obtusely pointed snout in lateral aspect, rounded canthal edges, concave loreal region, distinct tympanum and supratympanic fold, and medially webbed toes. Interorbital and internarial spaces concave or flat. Tympanum oval shaped. Snout and interorbital area smooth. Sides of head smooth or with glandular warts. Dorsum and upper flanks shagreened. Lower flanks granular. Dorsal parts of forelimbs have glandular warts. Thighs, shanks and feet dorsally smooth. Throat and chest smooth or shagreened. Belly and undersides of thighs granular. Dorsum and upper flanks have horn-like spinules in male. Resembles *P. folicola* but can be differentiated from it by obtusely pointed snout from lateral aspect, rounded canthal edges, thick orange or gold stripe that runs along canthal edge and thighs with posterior sides dark brown with white spots (vs rounded snout from lateral aspect, sharp canthal edges and absence of thick orange or gold stripe along canthal edge, and thighs with posterior sides dark brown/black with white spots). (Figs 1–3).

Colour Dorsum of body light to dark brown. Whitish spots scattered on dorsolateral area. Thin yellow line runs down on mid-dorsum from snout to vent. Canthal edge, edge of upper eyelid, and supratympanic fold orange or yellowish. Dorsum of limbs has brown cross-bars. Thighs dark brown with light brown spots. Posterior sides of thigh brown marbled with white. (Figs 1–3). Belly pale brown with brown mottling.

Habits Nocturnal species that can be seen

Fig. 1 Dorsolateral aspect, with thick stripe along canthal edge indicated by arrow

on low vegetation and boulders at night. By day it can be seen among leaf litter on the forest floor.

Habitat and Distribution Habitat generalist that can be seen in closed canopy rainforests and home gardens. Low-country wet zone species distributed at elevations of 30–650m, including at Kottawa, Hiyare, Kanneliya, Sinharaja, Udamaliboda, Runakanda, Dombagaskanda and Kosmulla.
Status Endemic.

IUCN Red List Category Near Threatened.

Fig. 2 Dorsolateral aspect, with horn-like spinules indicated by arrow

Fig. 3 Posterior sides of thighs brown marbled with white, indicated by arrow

STUART'S SHRUB FROG *Pseudophilautus stuarti*

(Stuartge pañduru mädiyã)

First described as *Philautus stuarti* by Madhava Meegaskumbura and Kelum Manamendra-Arachchi in 2005. Subsequently placed in the genus *Pseudophilautus*, and thus the current name is *Pseudophilautus stuarti*.

Size SVL 24–32mm.

Identification Features Small frog with stout body, dorsally flat head, rounded snout in lateral aspect, sharp canthal edges, concave loreal region, distinct tympanum and supratympanic fold, and medially webbed toes. Interorbital and internasal spaces flat. Supernumerary tubercles on palms and feet. Snout, interorbital space, and anterior and posterior dorsum have horny spinules; sides of head, and upper and lower flanks smooth. Dorsal parts of forelimbs, thighs, shanks and feet smooth. Throat, chest, belly and undersides of thighs granular. Resembles *P. viridis* and can be distinguished from it by sharp canthal edges (vs indistinct canthal edges in *P. viridis*). (Figs 1–3).

Colour light green to olive-brown with a few ashy-green, ashy-yellow or black dots. Loreal and tympanic regions and tympanum green. Canthal edges and supratympanic fold yellow. (Figs 1–3). Abdomen ashy flesh coloured.

Habits Commonly found on moss-covered rocks and logs at night. By day, occurs in tree-holes, and on leaf litter and *Selleginella* plants.

Habitat and Distribution Occurs in submontane forests and cardamom plantations in the Knuckles Hills at 1,000–1,400m, including in Corbett's Gap, Thangappuwa and Riverston.

Status Endemic.

IUCN Red List Category Critically Endangered.

Fig. 1 Dorsolateral aspect

Fig. 2 Dorsolateral aspect

Fig. 3 Dorsolateral aspect

SLENDER SHRUB FROG *Pseudophilautus tanu*

(Tanu pañduru mädiyã)

First described as *Philautus tanu* by Madhava Meegaskumbura, Kelum Manamendra-Arachchi and Rohan Pethiyagoda in 2009. Later placed in the genus *Pseudophilautus*, and the current name is *Pseudophilautus tanu*.

Size SVL 13–18mm.

Identification Features Small frog with short body, laterally convex head, obtusely pointed snout in dorsal and lateral aspects, rounded canthal edges, distinct tympanum and medially webbed toes. Loreal region, and interorbital and internasal spaces flat. Tympanum oval shaped. Supratympanic fold absent. Fingers without lateral dermal fringe. Supernumerary tubercles on toes and on feet. Dorsal and lateral parts of head and body shagreened. Upper and lower parts of flanks granular. Dorsal and lateral parts of upper arms, lower arms, thighs, shanks and feet smooth. Very narrow dermal fold on mid-dorsum, extending from tip of snout to vent. Ventral parts of throat, chest, abdomen, upper and lower arms, anterior thighs, shanks and feet granular. (Figs 1 & 2).

Colour Dorsum of head and body pale brown. Dark brown stripe about as wide as pupil extends backwards from snout, fading away on mid-flank. Sides of head dark brown. About eight dark brown stripes of varying width on dorsum. Ground colour of body creamy light brown. Narrow creamy-brown stripe extends from snout, over eye to flank. (Figs 1 & 2).

Habits Nocturnal and arboreal species that can be seen on shrubs at night. Usually calls from deep within shrubs, which makes it difficult to locate.

Habitat and Distribution Inhabits open shrub habitats and able to disperse freely through secondary-forest corridors and suitable anthropogenic habitats. Recorded at elevations of 20–600m, including at Beraliya, Kanneliya, Kitulgala, Pituwala and Kottawa.

Status Endemic.

IUCN Red List Category Endangered.

Fig. 1 Dorsolateral aspect

Fig. 2 Lateral aspect

STRIPED SNOUT SHRUB FROG *Pseudophilautus temporalis*

(Hombu thirethi panduru mädiya)

First described as *Ixalus temporalis* by Albert Günther in 1864. After several changes to the generic name (that is, to *Ixalus leucorhinus*, *Rhacophorus temporalis* and *Philautus temporalis*), it is currently known as *Pseudophilautus temporalis*. Known only from the lectotype deposited in the Natural History Museum, London, bearing the museum number 1947.2.6.9. There have been no sightings of the species since it was described in 1864.

Status Endemic.

IUCN Red List Category Extinct.

GÜNTHER'S SHRUB FROG *Pseudophilautus variabilis*

(Guntherge panduru mädiya)

First described as *Ixalus variabilis* by Albert Günther in 1858. After several changes to the generic name (that is, to *Philautus variabilis*, *Rhacophorus variabilis* and *Philautus variabilis*), it is currently known as *Pseudophilautus variabilis*. Known only from the lectotype deposited in the Natural History Museum, London, bearing the museum number 1947.2.7.87. Now considered to be extinct, as recent extensive field surveys have not been able to rediscover it.

Status Endemic.

IUCN Red List Category Extinct.

DULL GREEN SHRUB FROG *Pseudophilautus viridis*

(Anduru kola pañduru mädiyã)

First described as *Philautus viridis* by Kelum Manamendra-Arachchi and Rohan Pethiyagoda in 2005. Later placed in the genus *Pseudophilautus*, and thus its current name is *Pseudophilautus viridis*.

Size SVL 27–36mm.

Identification Features Small frog with stout body, dorsally convex head, blunt, rounded snout in lateral aspect, indistinct canthal edges, concave loreal region, distinct tympanum and supratympanic fold, and medially webbed toes. Interorbital and internarial spaces flat. Tympanum oval shaped. Lateral dermal fringes on fingers. Snout, interorbital area, sides of head, dorsum and upper flanks have horn-like spinules. Lower flanks granular. Dorsal part of forelimbs, thighs, shanks and feet shagreened. Throat, chest, belly and undersides of thighs granular. Nuptial pads absent in male. Resembles *P. stuarti* and can be distinguished from it by indistinct canthal edges (vs sharp canthal edges in *P. stuarti*). (Figs 1–3).

Colour Dorsum, dorsal head, loreal region, cheeks, tympanic region, tympanum and limbs dorsally light green olive-brown. Upper edge of supratympanic fold and outer edge of upper eyelid yellow. (Figs 1–3).

Habits Nocturnal species that perches on vegetation at 1–5m above the ground and on grass in open areas. By day, can be seen among the leaf litter of the forest floor and in tree-holes.

Habitat and Distribution Habitat generalist, found in both open and closed canopy vegetation, including cloud forests and adjacent anthropogenic habitats. Restricted to the Central Hills, being recorded at elevations of 1,400–1,830m, including at Agrabopath, Ambewela, Hakgala and Loolkandura. Inhabits submontane and montane forests, agricultural fields and pine plantations.

Status Endemic.

IUCN Red List Category Endangered.

Fig. 1 Dorsolateral aspect

Fig. 2 *Dorsolateral aspect*

Fig. 3 *Ventral aspect*

White-blotched Shrub Frog *Pseudophilautus zal*

(Sudu pulli panduru mädiya)

First described as *Philautus zal* by Kelum Manamendra-Arachchi and Rohan Pethiyagoda in 2005. Since the Sri Lankan and Indian *Philautus* was transferred to the genus *Pseudophilautus*, the current name is *Pseudophilautus zal*. Known only from the holotype (BMNH 1947.2.7.94) and two paratypes deposited at the Natural History Museum, London, collected in Sri Lanka before 1947. Since the species has not been seen since then, it is considered to be extinct.

Status Endemic.

IUCN Red List Category Extinct.

Rummassala Shrub Frog *Pseudophilautus zimmeri*

(Rummassala panduru mädiya)

First described as *Rhacophorus zimmeri* by Ernst Ahl in 1927. After changes to the generic name (that is, to *Rhacophorus zimmeri* and *Philautus zimmeri*), it is currently known as *Pseudophilautus zimmeri*. Known only from the holotype deposited in the Zoological Museum of Berlin, bearing the museum number ZMB 6111. There have been no sightings of the species since it was described in 1927, and it is thus now considered to be extinct.

Status Endemic.

IUCN Red List Category Extinct.

GANNORUWA SHRUB FROG *Pseudophilautus zorro*

(Gannoruwa pañduru mädiyã)

First described as *Philautus zorro* by Kelum Manamendra-Arachchi and Rohan Pethiyagoda in 2005. Sri Lankan and Indian *Philautus* species were later placed in the genus *Pseudophilautus*, and thus the current name is *Pseudophilautus zorro*.

Size SVL 22–30mm.

Identification Features Small frog with elongated body, dorsally concave head, pointed snout in lateral aspect, sharp canthal edges, flat interorbital space, calcars at heels, distinct tympanum and supratympanic region, and medially webbed toes. Loreal region and internarial space concave. Tympanum oval shaped. Snout, interorbital area and dorsum have horn-like spinules. Sides of head, upper flanks, thighs and shanks dorsally with glandular warts. Narrow dermal fringe on mid-dorsum from tip of snout to vent. Anterior dorsum has ')('-shaped pattern of tubercles. Dorsal parts of forelimbs, throat, mid-chest and feet smooth. Belly and lower flanks granular. Ventral sides of thighs smooth. Resembles *P. cuspis* and can be distinguished by supernumerary tubercles on feet, concave loreal region, dorsally concave head and internarial region (vs absence of supernumerary tubercles on feet, and presence of flat loreal region, dorsally convex head and flat internarial region in *P. cuspis*). (Figs 1–3).

Colour Dorsum beige to greyish-brown with black patches or pigments. Head laterally dark brown or black. Upper part of tympanum and loreal region black. Upper edge of supratympanic fold bright yellowish-light brown. Upper lip grey. Dorsal and lateral parts of forelimbs greyish-brown or dark brown. Thighs, shanks and feet grey with dark brown cross-bars. (Figs 1–3).

Habits Usually found on the forest floor where there is a thick leaf-litter layer and dense canopy cover. Often seen during the day.

Habitat and Distribution Habitat generalist,

Fig. 1 Dorsolateral aspect, with pointed snout indicated by arrow

inhabiting closed canopy forests, secondary forests and densely planted home gardens with a lot of leaf litter. Restricted to wet and intermediate zones at elevations of 500–800m. Known from Gannoruwa, Hantana, Udawattakelle, Kithulgala, Dotheloya, Katugasthota, Illukkumbura, Owilikanda and Nawalapitiya.

Status Endemic.

IUCN Red List Category Vulnerable.

Fig. 2 Dorsolateral aspect

Fig. 3 Ventral aspect

Genus *Taruga*

Taruga are comparatively small tree-frogs with a SVL of 35–45mm. They were previously placed in the genus *Polypedates*. However, morphological and molecular phylogenetic data suggested that they were unique and merited their own genus. They are characterized by a dorsolateral glandular fold that extends from the posterior margin of the upper eyelid to the mid-flanks, prominent calcars at the ends of the tibias, pointed snout and 6–10 prominent conical tubercles surrounding the cloaca. Like *Polypedates* species, *Taruga* species create foam nests on surfaces above water bodies. Once the eggs hatch the tadpoles fall into the water, where they live until metamorphosis. The genus is endemic to Sri Lanka, and comprises three species.

Mountain Tree-frog *Taruga eques*

(Kandukara pahimbu gas mädiya)

First described as *Polypedates eques* by Albert Günther in 1858. After several changes to the generic name (that is, to *Rhacophorus eques*, *R. cruciger eques* and *Polypedates eques*), it is currently known as *Taruga eques*.

Size SVL 32–71mm.

Identification Features Medium to large frog with elongated body, medially webbed toes, digits with distinct circular discs, long hindlimbs and forelimbs, pointed calcars at heels, dorsolateral fold from back of eye to mid-flank, conical tubercles around cloaca, elongated head, pointed and triangular snout when viewed dorsally, angular canthal edges, concave loreal region, rounded or vertically oval tympanum and distinct supratympanic fold. Both edges of digits have cutaneous fringes. Narrow glandular fold along tarsi and outer edge of fifth toe. Dorsum smooth and venter granular. Resembles *T. fastigo* and *T. longinasus*. Can be distinguished from *T. longinasus* by moderately elongated and pointed snout and hourglass mark on dorsum (vs elongated and pointed snout and absence of hourglass mark on dorsum in *T. longinasus*). Can be distinguished from *T. fastigo* by absence of black line (may be present as dotted line) on lower flank connecting axilla to groin (vs presence of black line on lower flank connecting axilla to groin in *T. fastigo*). (Figs 1–3).

Fig. 1 Dorsolateral aspect, with hourglass mark indicated by left-hand arrow, pointed calcar on heel by right-hand arrow

Colour Dorsal colour a mixture of grey, brown and orangish-yellow, and most individuals have characteristic hourglass mark on dorsum. Loreal region reddish-orange and tip of snout dark brown. Calcar reddish-orange. Dorsal sides of limbs have brown cross-bars. (Figs 1–3). Ventral side light yellow.

Habits Nocturnal and arboreal species. May be seen by day on the ground in grassy locations abutting water bodies. Amplexus takes place close to shallow pools and on grass.

Habitat and Distribution Occurs in both undisturbed forests and anthropogenic habitats such as home gardens and tea estates. Mainly found in grassland, marshes and low vegetation along forest edges, and close to aquatic habitats, in submontane and montane forests at elevations above 1,200m.

Status Endemic.

IUCN Red List Category Endangered.

Fig. 2 *Dorsolateral aspect*

Fig. 3 *Ventral aspect*

MORNINGSIDE SADDLED TREE-FROG *Taruga fastigo*

(Enasalwatta gas mädiyã)

First described as *Polypedates fastigo* by Kelum Manamendra-Arachchi and Rohan Pethiyagoda in 2001. Later placed in the genus *Taruga* and currently known as *Taruga fastigo*.

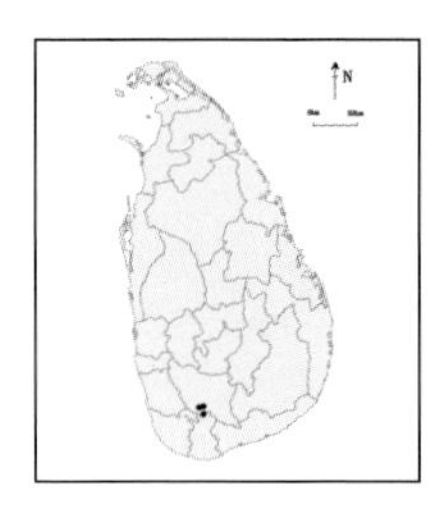

Size SVL 35–64mm.

Identification Features Medium to large frog with elongated body, medially webbed toes, digits with distinct circular discs, long hindlimbs and forelimbs, pointed calcars at heels, dorsolateral fold from back of eye to mid-flank, conical tubercles around cloaca, elongated head, pointed and triangular snout when viewed dorsally, angular and sharp canthal edges, concave loreal region, horizontally oval tympanum distinct and distinct supratympanic fold. Internarial area concave. Interorbital area flat. Skin of dorsum, chin and chest smooth. Granular on abdomen and undersides of femurs. Resembles *T. eques* and *T. longinasus*. Can be distinguished from *T. longinasus* by moderately elongated and pointed snout, and hourglass mark on dorsum (vs elongated and pointed snout, and absence of hourglass mark on dorsum in *T. longinasus*). Can be distinguished from *T. eques* by black line on lower flanks connecting axilla to groin (vs absence of black line (may be present as dotted line) on lower flanks connecting axilla to groin in *T. eques*). (Figs 1 & 2).

Colour Dorsum brown or olive-green. Incomplete dark brown hourglass-shaped marking on dorsum, its outline darker. Canthal ridge and supratympanic fold red. Edges of lips black. Venter white or pale yellow. Chin and gular area with or without black markings. Bases of upper arms have black oval markings. Calcar and tubercles around vent orange or yellow. (Figs 1 & 2).

Habits Nocturnal and arboreal species that can be seen close to water bodies at night. Tadpoles found in shallow, stagnant (<15cm-deep) pools with thick leaf-litter substrate in heavily shaded forests.

Habitat and Distribution Found in forests and cardamom plantations. Restricted to elevations above 1,000m in the Rakwana mountain range, and known from Morningside and Enasalwatte in the Sinharaja World Heritage Site.

Status Endemic.

IUCN Red List Category Critically Endangered.

Fig. 1 Dorsolateral aspect

Fig. 2 Dorsolateral aspect

LONG-SNOUT TREE-FROG *Taruga longinasus*

(Dick-hombu gas mädiyã)

First described as *Polypedates nasutus* by Albert Günther in 1869. After several name changes (to *Rhacophorus longinasus*, *R. (Rhacophorus) longinasus* and *Polypedates longinasus*), it is currently known as *Taruga longinasus*.

Size SVL 41–60mm.

Identification Features Medium-sized frog with elongated body, medially webbed toes, digits with distinct circular discs, long hindlimbs and forelimbs, pointed calcars at heels, dorsolateral fold from back of eye to mid-flank, conical tubercles around cloaca, elongated head, very pointed and triangular snout when viewed dorsally, angular canthal edges, grooved loreal region, round or horizontally oval tympanum and distinct supratympanic fold. Both edges of digits with cutaneous fringes. Narrow glandular fold along tarsi and outer edge of fifth toe. Dorsum, chin and chest smooth, venter finely granular. Lower sides of thighs have white tubercles. Resembles *T. fastigo* and *T. eques*, from which it can be distinguished by extremely elongated and pointed snout and absence of hourglass mark on dorsum (vs moderately elongated and pointed snout and hourglass mark on dorsum in *T. eques* and *T. fastigo*). (Figs 1 & 2).

Colour Dorsal colour varies from beige to dark brown. Bright red band from tip of snout to middle of flank. Lateral colour dark brown. Lips yellow or white. Limbs light brown or dark brown with cross-bars. Calcar orange. (Figs 1 & 2). Venter white or pale yellow.

Habits Usually found around well-shaded pools. During breeding season, 8–10 males gather around a pool and call from overhanging vegetation. In anthropogenic habitats, has been seen around abandoned wells at night.

Habitat and Distribution Occurs in both primary and secondary forests, and well-shaded anthropogenic habitats of wet zone at 150–1,000m above sea level. Known from Ambagamuwa, Kitulgala, Sinharaja, Hapugasthenna and Kanneliya.

Status Endemic.

IUCN Red List Category Endangered.

Fig. 1 Dorsolateral aspect

Fig. 2 *Lateral aspect, with very pointed snout indicated by arrow*

Fig. 3 *Ventral aspect*

ORDER Gymnophiona (Caecilians)
Caecilians are limbless, slender amphibians that resemble earthworms. Most are fossorial and live in moist soils close to streams, lakes and swamps. A few are aquatic. They are rarely seen due to their nocturnal behaviour and secretive nature. They have blunt, bullet-shaped heads and cylindrical, limbless bodies with short tails. The bodies are segmented by grooves (primary grooves). The blunt heads are used for burrowing into soil and they employ serpentine movement. Their eggs are fertilized inside the female's body and males have a copulatory organ (phallodeum) for transferring sperm to the female's reproductive tract. The offspring may develop internally or externally. Minute scales (dermal scales) are present in the skin. Their eyes are vestigial and lie beneath the skin or skull bones. They lack external ear openings. There is a short, retractile sensory tentacle on each side of the head between the eye and nostril, which is used to locate prey. Caecilians occur in the tropics worldwide except in Papua-Australia. Approximately 200 species in 10 families (Caecilidae, Chikilidae, Dermophiidae, Herpelidae, Ichthyophiidae, Rhinatrematidae, Scolecomorphidae, Siphonopidae, Typhlonectidae and Uraeotyphlidae) are known. The only family found in Sri Lanka is the Ichthyophiidae.

Family Ichthyophiidae (Asiatic Tailed Caecilians)
The Ichthyophiidae consists of 57 species that belong to two genera: *Ichthyophis* and *Uraeotyphlus*. Members of this family are found in India, Sri Lanka and Southeast Asia. Their primary annuli are divided by secondary and tertiary grooves. The body ends in a short true tail. The eyes are visible externally and lie in bony sockets beneath the skin. The sensory tentacles lie between the eyes and nostrils. Reproduction is indirect – that is, they have an aquatic larval stage. Three species representing the genus *Ichthyophis* are found in Sri Lanka.

Genus *Ichthyophis*
Consists of 50 medium to large (500mm) species distributed from Sri Lanka and India to Southeast Asia, characterized by primary annuli that are subdivided by secondary annuli. Three species are known from Sri Lanka.

COMMON YELLOW BAND CAECILIAN *Ichthyophis glutinosus*

(Kaha iridanda)

The first amphibian to be described from Sri Lanka, named *Caecilia glutinosa* by Carl Linnaeus in 1758. After several changes to the generic name (that is, to *Ichthyophis glutinosus*, *Epicrium glutinosum*, *Caecilia* (*Ichthyophis*) *glutinosa* and *Ichthyophis* (*Epicrium*) *glutinosus*), currently known as *Ichthyophis glutinosus*.

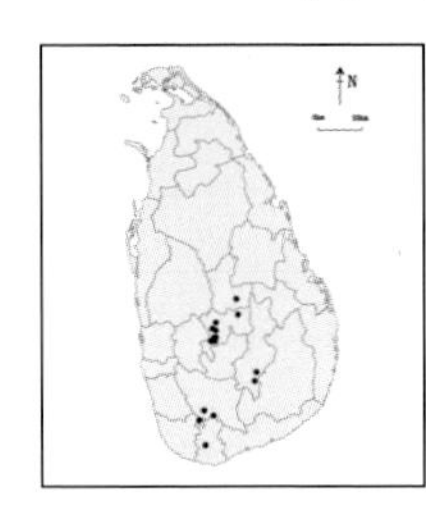

Size TL 165–410mm.

Identification Features Large caecilian with elongated, worm-like, limbless body, lateral yellow stripe running from head to tail, and 329–415 annuli. Body from head to tail more or less of the same girth and has a slimy appearance. Head oval and distinct from body. Eyes distinct. Posterior annuli transverse and posteriorly directed. Tail short. Resembles *I. pseudangularis*, from which it can be distinguished by 329–415 annuli on body (vs 269–304 annuli in *I. pseudangularis*). (Figs 1–3).

Colour Dorsal colour purplish-brown or blackish-brown and belly lighter. Distinct yellow stripe from head to tail on both sides of body. Stripe breaks at first collar and spotted at angle of jaw. Vent surrounded by pale patch. (Figs 1–3).

Habits Nocturnal, fossorial and burrowing

species that moves on land at night for foraging and other activities. Usually found among moist decaying leaf litter, under decaying logs, boulders and cow-dung heaps, and in cool, damp humus. Feeds voraciously on earthworms and soil-dwelling insect larvae. Female lays about 25 eggs in cavity in humus or under stones near streams, and coils around eggs and stays with them.

Habitat and Distribution Inhabits forests, paddy fields, vegetable gardens, tea plantations, banana groves and cattle sheds. The most widespread caecilian in Sri Lanka, being distributed in wet zone at about 150–1,400m elevation. Has been seen in Ambagamuwa, Gampola, Nawalapitiya, Wattegama, Corbet's Gap, Gammaduwa, Peradeniya, Panvilatenna, Gelioya, Knuckles, Hantana, Kandy, Runakanda, Namunukula, Passara, Dellawa and Kudawa (Sinharaja).

Status Endemic.

IUCN Red List Category Vulnerable.

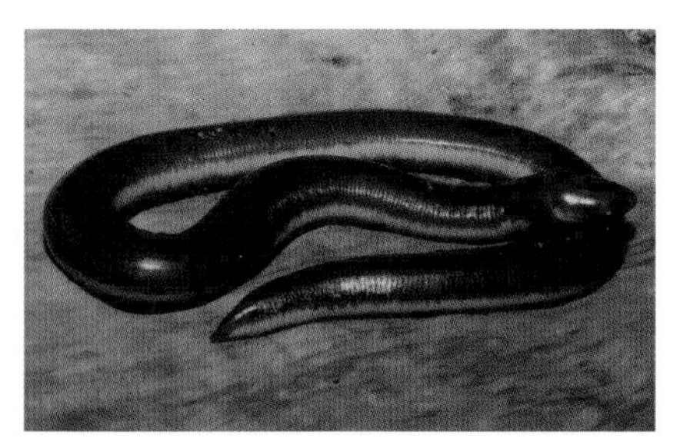

Fig. 1 Dorsolateral aspect

Fig. 2 Dorsolateral aspect

Fig. 3 Dorsolateral aspect

Brown Caecilian/Pattipola Caecilian *Ichthyophis orthoplicatus*

(Dumburu iridanda)

First described as *Ichthyophis orthoplicatus* by Edward Harrison Taylor in 1965. Later renamed as *Ichthyophis taprobanicensis* by E. H. Taylor, and currently known as *Ichthyophis orthoplicatus*.

Size TL 208–303mm.

Identification Features Comparatively small caecilian with, elongated, worm-like, limbless body and 290–335 annuli. Body from head to tail more or less of same girth and has slimy appearance. Lacks lateral stripe that runs along body. Head oval and distinct from body. Eyes distinct and tentacular opening is a horseshoe-shaped groove. Nostrils visible dorsally. Can be easily distinguished from the other two *Ichthyophis* species in Sri Lanka by absence of lateral stripe that runs along body (vs presence of yellow lateral stripe that runs along body in *I. glutinosus* and *I. pseudangularis*). (Figs 1 & 2).

Colour Dorsal colour light to dark brown and venter lighter. Tentacular opening lighter than background. Light ring around eye and white patch around vent. Tail has white patch. (Figs 1 & 2).

Habits Found under moist, decaying leaf litter, decaying logs and cool humus.

Habitat and Distribution Inhabits forests and tea estates. Distributed in mountains of south-central Sri Lanka in the Uva province at up to 1,890m elevation.

Status Endemic.

IUCN Red List Category Endangered.

Fig. 1 Dorsolateral aspect

Fig. 2 Dorsolateral aspect

Lesser Yellow-banded Caecilian *Ichthyophis pseudangularis*

(Kuda kaha iridanda)

First described as *Ichthyophis pseudangularis* by Edward Harrison Taylor in 1965.

Size TL 199–314mm.

Identification Features Medium-sized caecilian with elongated, worm-like, limbless body, lateral yellow stripe running from head to tail, and 269–304 annuli. Four or less annuli present behind cloaca. Body from head to tail more or less of same girth and has slimy appearance. Head oval and distinct from body. Head small, narrower than body, and bluntly pointed. Eyes, tentacular opening and nostrils distinct. Dorsal folds and grooves transverse, curving forwards on mid-line. Ventral folds angular, except posterior fifth of body. Resembles *I. glutinosus*, from which it can be distinguished by 269–304 annuli on body (vs 329–415 annuli in *I. glutinosus*). (Figs 2 & 3).

Colour Dorsal colour maroonish-brown to dark blackish-brown, and belly light. Distinct yellow stripe runs from head to tail on both sides of body; this narrows or is broken around neck region (Figs 2 & 3).

Habits Nocturnal, fossorial and burrowing species that moves on land at night for foraging and other activities. Feeds voraciously on earthworms. Female lays about 25 eggs in cavity in humus or under stones near streams, and coils around eggs and stays with them.

Habitat and Distribution Found in forests and tea estates. Confined to south-west wet zone at about 250–1,200m above sea level. Known from Watawala, Ratnapura, Akuressa, Enasalwatta-Sinharaja, Deniyaya, Balangoda and Hiyare (Galle).

Status Endemic.

IUCN Red List Category Vulnerable.

Fig 1 Head detail

Fig. 2 Dorsolateral aspect

Fig. 3 Dorsolateral aspect

Checklist of the Amphibians of Sri Lanka

(Follows nomenclature available as of May 2021)

CR Critically Endangered
EN Endangered
VU Vulnerable
NT Near Threatened
LC Least Concern
DD Data Deficient
EX Extinct
NE Not Evaluated

Scientific name	English name	Status	Conservation Status
Bufonidae (Toads)			
Adenomus kandianus	Kandiyan Dwarf Toad	Endemic	EN
Adenomus kelaartii	Kelaart's Dwarf Toad	Endemic	VU
Duttaphrynus kotagamai	Kotagama's Toad	Endemic	EN
Duttaphrynus melanostictus	Common Indian Toad	Indigenous	LC
Duttaphrynus noellerti	Noellert's Toad	Endemic	CR
Duttaphrynus scaber	Ferguson's Toad	Indigenous	LC
Dicroglossidae (Fork-tongued Frogs)			
Euphlyctis cyanophlyctis	Indian Skipper Frog	Indigenous	LC
Euphlyctis hexadactylus	Six-toe Green Frog	Indigenous	LC
Hoplobatrachus crassus	Jerdon's Bull Frog	Indigenous	LC
Hoplobatrachus tigerinus	Indian Bull Frog	Indigenous	LC
Minervarya agricola	Common Paddy Field Frog	Indigenous	LC
Minervarya greenii	Montane Paddy Field Frog	Endemic	EN
Minervarya kirtisinghei	Sri Lanka Paddy Field Frog	Endemic	NT
Nannophrys ceylonensis	Sri Lanka Rock Frog	Endemic	VU
Nannophrys guentheri	Guenther's Rock Frog	Endemic	EX
Nannophrys marmorata	Kirtisinghe's Rock Frog	Endemic	EN
Nannophrys naeyakai	Sri Lanka Tribal Rock Frog	Endemic	EN
Sphaerotheca pluvialis	Jerdon's Sand Frog	Indigenous	LC
Sphaerotheca rolandae	Marbled Sand Frog	Indigenous	LC
Microhylidae (Narrow-mouth Frogs)			
Microhyla karunaratnei	Karunaratne's Narrow-mouth Frog	Endemic	EN
Microhyla mihintalei	Mihintale Red Narrow-mouth Frog	Endemic	LC
Microhyla ornata	Ornate Narrow-mouth Frog	Indigenous	LC
Microhyla zeylanica	Sri Lanka Narrow-mouth Frog	Endemic	EN
Uperodon nagaoi	Nagao's Pug Snout Frog	Endemic	EN
Uperodon obscurus	Brown Pug Snout Frog	Endemic	NT
Uperodon palmatus	Half-webbed Pug Snout Frog	Endemic	EN
Uperodon rohani	Rohan's Pug Snout Frog	Endemic	LC
Uperodon systoma	Balloon Frog	Indigenous	LC
Uperodon taprobanicus	Common Bull Frog	Indigenous	LC
Nyctibatrachidae (Wrinkled Frogs)			
Lankanectes corrugatus	Corrugated Water Frog	Endemic	NT
Lankanectes pera	Knuckles Corrugated Water Frog	Endemic	CR
Ranidae (True Frogs)			
Hydrophylax gracilis	Gravenhorst's Golden-backed Frog	Endemic	LC
Indosylvirana serendipi	Sri Lankan Golden-backed Frog	Endemic	VU
Indosylvirana temporalis	Günther's Golden-backed Frog	Endemic	VU
Rhacophoridae (Afro-Asian Tree-frogs)			
Polypedates cruciger	Common Hour-glass Tree-frog	Endemic	LC

Scientific name	English name	Status	Conservation Status
Polypedates maculatus	Spotted Tree-frog	Indigenous	LC
Polypedates ranwellai	Ranwella's Tree-frog	Endemic	EN
Pseudophilautus abundus	Labugagama Shrub Frog	Endemic	LC
Pseudophilautus adspersus	Thwaites's Shrub Frog	Endemic	EX
Pseudophilautus alto	Horton Plains Shrub Frog	Endemic	EN
Pseudophilautus asankai	Asanka's Shrub Frog	Endemic	EN
Pseudophilautus auratus	Golden Shrub Frog	Endemic	EN
Pseudophilautus bambaradeniyai	Bambaradeniya's Shrub Frog	Endemic	CR
Pseudophilautus caeruleus	Blue-thigh Shrub Frog	Endemic	EN
Pseudophilautus cavirostris	Hollow-snouted Shrub Frog	Endemic	VU
Pseudophilautus conniffae	Conniff's Shrub Frog	Endemic	EN
Pseudophilautus cuspis	Sharp-snouted Shrub Frog	Endemic	EN
Pseudophilautus dayawansai	Dayawansa's Shrub Frog	Endemic	CR
Pseudophilautus decoris	Elegant Shrub Frog	Endemic	CR
Pseudophilautus dilmah	Dilmah Shrub Frog	Endemic	NE
Pseudophilautus dimbullae	Dimbulla Shrub Frog	Endemic	EX
Pseudophilautus eximius	Queenwood Shrub Frog	Endemic	EX
Pseudophilautus extirpo	Blunt-snouted Shrub Frog	Endemic	EX
Pseudophilautus femoralis	Leaf-nesting Shrub Frog	Endemic	EN
Pseudophilautus fergusonianus	Ferguson's Shrub Frog	Endemic	LC
Pseudophilautus folicola	Leaf-dwelling Shrub Frog	Endemic	VU
Pseudophilautus frankenbergi	Frankenberg's Shrub Frog	Endemic	EN
Pseudophilautus fulvus	Knuckles Shrub Frog	Endemic	EN
Pseudophilautus hallidayi	Halliday's Shrub Frog	Endemic	VU
Pseudophilautus halyi	Pattipola Shrub Frog	Endemic	EX
Pseudophilautus hankeni	Hanken's Shrub Frog	Endemic	EN
Pseudophilautus hypomelas	Webless Shrub Frog	Endemic	EN
Pseudophilautus hoffmanni	Hoffman Shrub Frog	Endemic	NE
Pseudophilautus hoipolloi	Anthropogenic Shrub Frog	Endemic	NE
Pseudophilautus jagathgunawardanai	Jagath Gunawardana's Shrub Frog	Endemic	CR
Pseudophilautus karunarathnai	Karunarathna's Shrub Frog	Endemic	CR
Pseudophilautus leucorhinus	White-nosed Shrub Frog	Endemic	EX
Pseudophilautus limbus	Haycock Shrub Frog	Endemic	EN
Pseudophilautus lunatus	Handapan Ella Shrub Frog	Endemic	CR
Pseudophilautus macropus	Bigfoot Shrub Frog	Endemic	VU
Pseudophilautus maia	Good Mother Shrub Frog	Endemic	EX
Pseudophilautus malcolmsmithi	Malcomsmith's Shrub Frog	Endemic	EX
Pseudophilautus microtympanum	Small-eared Shrub Frog	Endemic	EN
Pseudophilautus mittermeieri	Mittermeier's Shrub Frog	Endemic	VU
Pseudophilautus mooreorum	Moore's Shrub Frog	Endemic	CR
Pseudophilautus nanus	Southern Shrub Frog	Endemic	EX
Pseudophilautus nasutus	Pointed-snout Shrub Frog	Endemic	EX
Pseudophilautus nemus	Whistling Shrub Frog	Endemic	EN
Pseudophilautus newtonjayawardanei	Newton Jayawardane's Shrub Frog	Endemic	CR
Pseudophilautus ocularis	Golden-eye Shrub Frog	Endemic	CR
Pseudophilautus oxyrhynchus	Sharp-snouted Shrub Frog	Endemic	EX
Pseudophilautus papillosus	Papillated Shrub Frog	Endemic	NE
Pseudophilautus pardus	Leopard Shrub Frog	Endemic	EX
Pseudophilautus pleurotaenia	Side-striped Shrub Frog	Endemic	VU
Pseudophilautus poppiae	Poppy's Shrub Frog	Endemic	CR
Pseudophilautus popularis	Common Shrub Frog	Endemic	NT
Pseudophilautus procax	Cheeky Shrub Frog	Endemic	CR
Pseudophilautus puranappu	Puran Appu's Shrub Frog	Endemic	CR
Pseudophilautus regius	Polonnaruwa's Shrub Frog	Endemic	VU
Pseudophilautus reticulatus	Reticulated Thigh Shrub Frog	Endemic	VU
Pseudophilautus rugatus	Farnland Shrub Frog	Endemic	EX
Pseudophilautus rus	Kandiyan Shrub Frog	Endemic	NT

Scientific name	English name	Status	Conservation Status
Pseudophilautus samarakoon	Samrakoon's Shrub Frog	Endemic	CR
Pseudophilautus sarasinorum	Muller's Shrub Frog	Endemic	EN
Pseudophilautus schmarda	Schmarda's Shrub Frog	Endemic	EN
Pseudophilautus schneideri	Schneider's Shrub Frog	Endemic	VU
Pseudophilautus semiruber	Annandale's Shrub Frog	Endemic	EN
Pseudophilautus silus	Pug-nosed Shrub Frog	Endemic	EN
Pseudophilautus silvaticus	Forest Shrub Frog	Endemic	EN
Pseudophilautus simba	Sinharaja Shrub Frog	Endemic	CR
Pseudophilautus singu	Sri Lanka Short-horned Shrub Frog	Endemic	EN
Pseudophilautus sirilwijesundarai	Siril Wijesundara's Shrub Frog	Endemic	CR
Pseudophilautus sordidus	Grubby Shrub Frog	Endemic	NE
Pseudophilautus steineri	Steiner's Shrub Frog	Endemic	EN
Pseudophilautus stellatus	Spotted Shrub Frog	Endemic	CR
Pseudophilautus stictomerus	Orange Canthal Shrub Frog	Endemic	VU
Pseudophilautus stuarti	Stuart's Shrub Frog	Endemic	CR
Pseudophilautus tanu	Slender Shrub Frog	Endemic	EN
Pseudophilautus temporalis	Striped-snout Shrub Frog	Endemic	EX
Pseudophilautus variabilis	Gunther's Shrub Frog	Endemic	EX
Pseudophilautus viridis	Dull Green Shrub Frog	Endemic	EN
Pseudophilautus zal	White-bloched Shrub Frog	Endemic	EX
Pseudophilautus zimmeri	Rumassala Shrub Frog	Endemic	EX
Pseudophilautus zorro	Gannoruwa Shrub Frog	Endemic	VU
Taruga eques	Mountain Saddled Tree-frog	Endemic	EN
Taruga fastigo	Morningside Saddled Tree-frog	Endemic	EN
Taruga longinasus	Long-snouted Tree-frog	Endemic	EN
Ichthyophiidae (Asiatic Tailed Caecilians)			
Ichthyophis glutinosus	Common Yellow Band Cecilian	Endemic	VU
Ichthyophis orthoplicatus	Brown Cecilian	Endemic	EN
Ichthyophis pseudangularis	Lesser Yellow-banded Cecilian	Endemic	VU

Further Information

Observing & Photographing Amphibians

Observing amphibians and recording their behaviour and natural history is an interesting hobby – and information based on such observations can be useful in science and also for the purposes of conservation. You can start by recording the presence and behaviour of amphibians in a home garden, and this can be further extended to their trips away from home. Record such observations in a notebook, along with the date and time of observation, prevailing weather conditions, habitat and particular behaviour. The observed amphibians can be photographed as a hobby or for later confirmation of the species identity. Clear, sharp images from several aspects of the amphibian, with a scale to determine the size, will be highly beneficial in later identification of the species.

The ideal equipment for photography is a digital SLR camera fitted with a macro lens (ideally a 100mm lens), and a flash. However, some modern mobile phone cameras may have the ability to take clear, sharp images. During photography, avoid keeping any frogs or tadpoles in direct sunlight as this could quickly overheat or dehydrate them and kill them. Higher lens apertures (11–16) and medium ISO sensitivities (100–400) with a fill-in flash in a DSLR camera will give you the best results in photography.

References

The following is a list of some sources that the authors referred to during the writing of this book. It is by no means a comprehensive list of literature on the subject – but rather a selection that we believe has made significant contributions to the understanding of Sri Lanka's amphibians, and to amphibians in general in relation to Sri Lankan species.

Adams, M. J. 1999. Correlated factors in amphibian declines: exotic species and habitat change in western Washington. *Journal of Wildlife Management* 63: 1162–1171.

AmphibiaWeb. 2020. <https://amphibiaweb.org> University of California, Berkeley, CA, USA.

Bahir, M. M., Meegaskumbura, M., Pethiyagoda, R., Manamendra-Arachchi, K. & C. Schneider. 2005. Reproduction and terrestrial direct development in Sri Lankan shrub frogs (Ranidae: Rhacophorinae: Philautus). *The Raffles Bulletin of Zoology*. Supplement No. 12: 339–350.

Batuwita, S., De Silva, M. & S. Udugampala. 2019. Description of a new species of *Pseudophilautus* (Amphibia: Rhacophoridae) from southern Sri Lanka. *Journal of Threatened Taxa* 11: 13120–13131.

Berger, L., Speare, R., Daszak, P., Green, D. E., Cunningham, A. A., Goggin, C. L., Slocombe, R., Ragan, M. A., Hyatt, A. D., Mcdonald, K. R., Hines, H. B., Lips, K. R., Marantelli, G. & H. Parkes. 1998. Chytridiomycosis causes amphibian mortality associated with population declines in the rain forests of Australia and Central America. *Proceedings of the National Academy of Science*, USA 95: 9031–9036.

Biju, S. D., Garg, S., Mahony, S., Wijayathilaka, N., Senevirathne, G. & M. Meegaskumbura. 2014. DNA barcoding, phylogeny and systematics of Golden-backed frogs (*Hylarana*, Ranidae) of the Western Ghats-Sri Lanka biodiversity hotspot, with the description of seven new species. *Contributions to Zoology* 83: 269–335.

Bossuyt, F. & A. Dubois. 2001. A review of the frog genus *Philautus* Gistel, 1848 (amphibia, anura, ranidae, rhacophorinae). *Zeylanica* 6: 1–112.

Boulenger, G. A. 1892. *Catalogue of the Batrachia Salientia s. Ecaudata in the Collection of the British Museum*. 2nd edn. Natural History Museum Publications, London, UK.

Brooks, T. M., Mittermeier, R. A., Mittermeier, C. G., Da Fonseca, G.A., Rylands, A. B., Konstant, W. R., Flick, P., Pilgrim, J., Oldfield, S., Magin, G. & C. Hilton-Taylor. 2002. Habitat loss and extinction in the hotspots of biodiversity. *Conservation Biology* 16: 909–923.

Bucciarelli, G. M., Clark, M. A., Delaney, K. S., Riley, S. P., Shaffer, H. B., Fisher, R. N., Honeycutt, R. L. & L. B. Kats. 2020. Amphibian responses in the aftermath of extreme climate events. *Scientific reports* 10: 1–7.

Clarke B. T. 1983. A morphological re-examination of the frog genus *Nannophrys* (Anura: Ranidae) with comments on its biology, distribution and relationships. *Zoological Journal of the Linnaean Society* 79: 377–398.

Daniels, R. J. R. 2005. *Amphibians of Peninsular India*. University Press, Hyderabad, India.

Daszak, P., Cunningham, A. A. & A. D. Hyatt. 2003. Infectious disease and amphibian population declines. *Diversity and Distributions* 9: 141–150.

de Silva, A. 1998. Plants used as snake repellents and in the traditional management of snakebite in Sri Lanka. *Ayurveda Sameekshawa* 1: 109–113.

de Silva, A. 2001. Some aquatic insects: predators of anuran larvae at Horton Plains National Park. *The Amphibia of Sri Lanka: Recent Research. Lyriocephalus (Special Issue)* 4: 145–146.

de Silva, A. 2001. Folklore, traditions and proverbs about amphibians in Sri Lanka. *The Amphibia of Sri Lanka: Recent Research. Lyriocephalus (Special Issue)* 4: 168–170.

de Silva, A. 2007. *The Diversity of Horton Plains*. Vijitha Yapa Publications, Colombo, Sri Lanka.

de Silva, A. & D. Mahaulpatha. 2007. *A Manual on Field Techniques on Herpetology for Sri Lanka*. Sri Lanka Association for the Advancement of Science & Ministry of Environment & Natural Resources, Colombo, Sri Lanka.

de Silva, A. 2009. *Amphibians of Sri Lanka: A Photographic Guide to Common Frogs, Toads and Caecilians*. Amphibia and Reptile Research Organization of Sri Lanka. Gampola, Sri Lanka.

de Silva, A. 2009. *The Incidence and Pattern of Malformations, Abnormalities, Injuries, and Parasitic Infection of Amphibians in Sri Lanka* (Preliminary Findings). Final Report: Amphibian Specialists Group, Seed Grant.

de Silva, A. & K. Ukuwela. 2020. *A Naturalists Guide to the Reptiles of Sri Lanka*. 2nd edn. John Beaufoy Publishing Ltd., Oxford, UK.

de Silva, C. M. A & K. D. C. Wickramaratne. 1956. *Treatment for Snakebite*. Ceylon National Museums Manuscript Series. Vol. VIII, Govt Press, Colombo, Ceylon.

Dutta, S. K. & K. Manamendra-arachchi. 1996. *The Amphibian Fauna of Sri Lanka*. Wildlife Heritage Trust of Sri Lanka. Colombo, Sri Lanka.

Frost, D. 1998. Amphibian species of the world < https://amphibiansoftheworld.amnh.org/>. American Museum of Natural History, New York, USA.

Frost, D. R., Grant, T., Faivovich, J., Bain, R. H., Haas, A., Haddad, C. F. B., de Sá, R. O., Channing, A., Wilkinson, M., Donnellan, S. C., Raxworthy, C. J., Campbell, J. A., Blotto, B. L., Moler, P. E., Drewes, R. C., Nussbaum, R. A., Lynch, J. D., Green, D. M. & W. C. Wheeler. 2006. The amphibian tree of life. *Bulletin of American Museum Natural History* 297: 1–370.

Garg, S., Senevirathne, G., Wijayathilaka, N., Phuge, S., Deuti, K., Manamendra-Arachchi, K., Meegaskumbura, M. & S. D. Biju. 2018. An integrative taxonomic review of the South Asian microhylid genus *Uperodon*. *Zootaxa* 4384: 001–088.

Geiger, W. 1960. *The Mahavamsa*. Government Publications Bureau, Colombo. Sri Lanka.

Gower, D. J., Bahir, M. M., Mapatuna, Y., Pethiyagoda, R., Raheem, D. & M. Wilkinson. 2005. Molecular phylogenetics of Sri Lankan *Ichthyophis* (Amphibia: Gymnophiona: Ichthyophiidae), with discovery of a cryptic species. In: Yeo, D. C. J., Ng, P. K. L. & R. Pethiyagoda (eds), Contributions to biodiversity exploration and research in Sri Lanka. *The Raffles Bulletin of Zoology (Supplement No. 12)*: 153–161.

Gunasena, D. 1963. *Visha vaidiya prkaranaya*. Modern Book Co. Nugegoda, Sri Lanka.

Gunatilleke, I. A. U. N. & C. V. S. Gunatilleke. 1990. Problems of Deforestation and Forest Conservation, (paper presented at) *Natural Resource Conservation, Ceylon Social Institute*, Kandy, Sri Lanka.

Gunther, A. 1858. *Catalogue of the Batrachia Salientia in the Collection of the British Museum, London*. Trustees of the British Museum London, London, UK.

Henry, P. F. P. 2000. Aspects of amphibian anatomy and physiology. In: Sparling, D. W., Linder, G. & C. A. Bishop (eds), *Ecotoxicology of amphibians and reptiles*. Society of Environmental Toxicology and Chemistry. pp. 1–110.

IUCN 2006. Global Amphibian Assessment. Available: www.globalamphibians.org.

IUCN Sri Lanka and Ministry of Environment and Natural Resources. 2007. *The 2007 Red List of Threatened Fauna and Flora of Sri Lanka*. Colombo, Sri Lanka.

IUCN Red List (2020) *Red List of the Threatened Amphibians of Sri Lanka*.

Jayawardena, B., Senevirathne, G., Wijayathilaka, N., Ukuwela, K., Manamendra-Arachchi, K. & M. Meegaskumbura. 2017. Species boundaries, biogeography and evolutionarily significant units in dwarf toads: *Duttaphrynus scaber* and *D. atukoralei* (Bufonidae: Adenominae). *Ceylon Journal of Science (Special Issue)* 46: 79–87.

Jayawardena, U. A., Navaratne, A. N., Amerasinghe, P. H. & R. S. Rajakaruna. 2011. Acute and chronic toxicity of four commonly used agricultural pesticides on the Asian common toad, *Bufo melanostictus* Schneider. *Journal of the National Science Foundation of Sri Lanka* 39: 267–276.

Karunarathna, S., Ranwala, S., Surasinghe, T. & M. Madawala. 2017. Impact of vehicular traffic on vertebrate fauna in Horton plains and Yala national parks of Sri Lanka: some implications for conservation and management. *Journal of Threatened Taxa* 9: 9928–993.

Kelaart, E. F. 1852. *Prodromus faunae zeylanicae: being contributions to the zoology of Ceylon*. Published by the author, Colombo, Ceylon.

Kirtisinghe, P. 1957. *The Amphibia of Ceylon*. Published by the author, Colombo, Ceylon.

Li, J.-T., Che, J., Murphy, R. W., Zhao, H., Zhao, E.-M., Rao, D.-Q. & Y.-P. Zhang. 2009. New insights to the molecular phylogenetics and generic assessment in the Rhacophoridae (Amphibia: Anura) based on five nuclear and three mitochondrial genes, with comments on the evolution of reproduction. *Molecular Phylogenetics and Evolution* 53: 509–522.

Manamendra-Arachchi, K. & R. Pethiyagoda. 1998. A synopsis of the Sri Lankan Bufonidae (Amphibia: Anura), with description of two new species. *Journal of South Asian Natural History* 3: 213–246.

Manamendra-Arachchi, K. & R. Pethiyagoda. 2005. The Sri Lankan shrub-frogs of the genus

Philautus Gistel, 1848 (Ranidae: Rhacophorinae), with description of 27 new species. In: Yeo, D. C. J., Ng, P. K. L. & R. Pethiyagoda (eds), Contributions to biodiversity exploration and research in Sri Lanka. *The Raffles Bulletin of Zoology (Supplement No. 12)*: 163–303.

Manamendra-Arachchi, K. & R. Pethiyagoda. 2006. *Sri Lankawe Ubhayajeeveen* [*'The amphibian fauna of Sri Lanka'*]. WHT Publishers, Colombo, Sri Lanka.

McMenamin, S. K., Hadly, E. A. & C. K. Wright 2008. Climatic change and wetland desiccation cause amphibian decline in Yellowstone National Park. *Proceedings of the National Academy of Sciences* 105: 16988–16993.

McNeely, J. A., Miller, K. R., Reid, W. V., Mittermeier, R. A. & T. B. Werner. 1990. *Conserving the World's Biological Diversity*. IUCN, WRI, CI & WWF US, Gland, Switzerland.

Meegaskumbura, M., Bossuyt, F., Pethiyagoda, R., Manamendra-Arachchi, K., Bahir, M., Milinkovitch, M. & C. Schneider. 2002. Sri Lanka: an amphibian hot spot. *Science* 298: 379–379.

Meegaskumbura, M. & K. Manamendra-Arachchi. 2005. Description of eight new species of shrub frogs (Ranidae: Rhacophorinae: *Philautus*) from Sri Lanka. In: Yeo, D. C. J., Ng, P. K. L. & R. Pethiyagoda (eds), Contributions to biodiversity exploration and research in Sri Lanka. *The Raffles Bulletin of Zoology (Supplement No. 12)*: 305–338.

Meegaskumbura, M., Manamendra-Arachchi, K., Schneider, C. J. & R. Pethiyagoda. 2007. New species amongst Sri Lanka's extinct shrub frogs (Amphibia: Rhacophoridae: *Philautus*). *Zootaxa* 1397: 1–15.

Meegaskumbura, M., Meegaskumbura, S., Bowatte, G., Manamendra-Arachchi, K., Pethiyagoda, R., Hanken, J. & C. J. Schneider. 2010. *Taruga* (Anura: Rhacophoridae), a new genus of foam-nesting tree frogs endemic to Sri Lanka. *Ceylon Journal of Science* (Bio. Sci.) 39: 75–94.

Meegaskumbura, M., Manamendra-Arachchi, K., Bowatte G. & S. Meegaskumbura. 2012. Rediscovery of *Pseudophilautus semiruber*, a diminutive shrub frog (Rhacophoridae: *Pseudophilautus*) from Sri Lanka. *Zootaxa* 3229: 58–68.

Meegaskumbura, M., Senevirathne, G., Wijayathilaka, N., Jayawardena, B., Bandara, C., Manamendra-Arachchi, K. & R. Pethiyagoda. 2015. The Sri Lankan torrent toads (Bufonidae: Adenominae: Adenomus): species boundaries assessed using multiple criteria. *Zootaxa* 3911: 245–261.

Meegaskumbura, M., Senevirathne, G., Manamendra-Arachchi, K., Pethiyagoda, R., Hanken, J. & C. J. Schneider. 2019. Diversification of shrub frogs (Rhacophoridae, *Pseudophilautus*) in Sri Lanka – timing and geographic context. *Molecular Phylogenetics and Evolution* 132: 14–24.

Mendelson, J. R., Lips, K. R., Gagliardo, R. W., Rabb, G. B., Collins, J. P., Diffendorfer, J. E., Daszak, P., Ibanez, D. R., Zippel, K. C., Lawson, D. P., Wright, K. M., Stuart, S. N., Gascon, C., da Silva, H. R., Burrowes, P. A., Joglar, R. L., la Marca, E., Lotters, S., du Preez, L. H., Weldon, C., Hyatt, A., Rodriguezmahecha, J. V., Hunt, S., Robertson, H., Lock, B., Raxworthy, C. J., Frost, D. R., Lacy, R. C., Alford, R. A., Campbell, J. A., Parra-Olea, G., Bolanos, F., Domingo, J. J. C., Halliday, T., Murphy, J. B., Wake, M. H., Coloma, L. A., Kuzmin, S. L., Price, M. S., Howell, K. M., Lau, M., Pethiyagoda, R., Boone, M., Lannoo, M. J., Blaustein, A. R., Dobson, A., Griffiths, R. A. L., Crump, M., Wake, D. B. & E. D. Brodie, Jr. 2006. Confronting amphibian declines and extinctions. *Science* 313: 48.

Naveendrakumar, G., Vithanage, M., Kwon, H. H., Iqbal, M. C. M., Pathmarajah, S. & J. Obeysekera. 2018. Five decadal trends in averages and extremes of rainfall and temperature in Sri Lanka. *Advances in Meteorology*, 2018: 1–13. eID 4217917.

Oliver, L. A., Prendini, E., Kraus, F. & C. J. Raxworthy. 2015. Systematics and biogeography of the *Hylarana* frog (Anura: Ranidae) radiation across tropical Australasia, Southeast Asia, and Africa. *Molecular Phylogenetics and Evolution* 90: 176–192.

Peloso, P. L. V., Frost, D. R., Richards, S. J., Rodrigues, M. T., Donnellan, S., Matsui, M., Raxworthy, C. J., Biju, S. D., Lemmon, E. M., Lemmon, A. R. & W. C. Wheeler. 2015. The impact of anchored phylogenomics and taxon sampling on phylogenetic inference in narrow mouthed frogs (Anura, Microhylidae). *Cladistics* 32: 113–140.

Pethiyagoda, R., Manamendra-Arachchi, K., Bahir, M. M. & M. Meegaskumbura. 2006. Sri Lankan Amphibians: Diversity, Uniqueness and Conservation. In: Bambaradeniya C. N.

B. (ed.), *The Fauna of Sri Lanka: Status of Taxonomy, Research and Conservation*, pp. 125–133. World Conservation Union in Sri Lanka, Colombo, Sri Lanka.

Pounds, J. A. & M. L. Crump. 1994. Amphibian declines and climate disturbance: the case of the Golden Toad and the Harlequin Frog. *Conservation Biology* 8:72–85.

Pounds, J. A., Fogden, M. P. L. & J. H. Campbell. 1999. Biological response to climate change on a tropical mountain. *Nature* 398: 611–615.

Premathilake, R. 2005a. Earliest agricultural land of the Horton Plains, central Sri Lanka. *Prajasakthi Journal*, Village Development, Training and Research Center, Sri Lanka. 24–28pp. (Text in Sinhalese).

Premathilake, R. 2005b. Emergence and development of prehistoric agriculture in central Sri Lanka. *History and Archaeology* Vol. I. Central Cultural Fund, Colombo, Sri Lanka.

Punchihewa, G. G. 2001. Frogs in antiquity and legend. *The Amphibia of Sri Lanka: Ongoing Research. Lyriocephalus (Special Issue)* 4: 171.

Rajakaruna, R. S., Samarawickrama, V. A. M. P. K. & K. B. Ranawana. 2007. Amphibian declines and possible etiologies: the case for Sri Lanka. *Journal of the National Science Foundation of Sri Lanka* 35: 3–8.

Rajakaruna, R. S., Piyatissa, P. M. J. R., Jayawardena, U. A., Navaratne, A. N. & P. H. Amerasinghe. 2008. Trematode infection induced malformations in the common hourglass treefrogs. *Journal of Zoology* 275: 89–95.

Rajapaksa, C. W. 1962. *Visha ausada snagrahaya*. Chitra Printers, Ambalangoda. 22pp.

Reaser, J. K. 2000. Amphibian declines: an issue overview. Federal Taskforce on Amphibian Declines and Deformities. Washington, DC, USA.

Relyea, R. A., Schoeppner, N. M. & J. T. Hoverman. 2005. Pesticides and amphibians: the importance of community context. *Ecological Applications* 15: 1125–1134.

Richman, A. D., Case, T. J. & T. D. Schwaner. 1988. Natural and unnatural extinction rates of reptiles on islands. *American Naturalist* 131: 611–630.

Robert, E. 1919. Native remedies used in snakebite etc. H. W. Cave & Co. Printers, Colombo, Ceylon.

Sanchez, E., Biju, S. D., Islam, M. M., Hasan, M. K., Ohler, A., Vences, M. & A. Kurabayashi. 2018. Phylogeny and classification of fejervaryan frogs (Anura: Dicroglossidae). *Salamandra* 54: 109–116.

Senevirathne, G. & M. Meegaskumbura. 2015. Life among crevices: osteology of *Nannophrys marmorata* (Anura: Dicroglossidae). *Zootaxa* 4032: 241–245.

Senevirathne, G., Samarawikrama, V. A. M. P. K., Wijayathilaka, N., Manamendra-Arachchi, K., Bowatte, G., Samarawikrama, D. R. N. S. & M. Meegaskumbura. 2018. A new frog species from rapidly dwindling cloud forest streams of Sri Lanka-*Lankanectes pera* (Anura, Nyctibatrachidae). *Zootaxa* 4461: 519–538.

Senaviratna, J. M. 1936. *Directory of Proverbs of the Sinhalese*. Times of Ceylon Co. Ltd. Colombo, Sri Lanka.

Seyone, K. N. V. 1998. *Some Old Coins Found in Early Ceylon (Sri Lanka)*. Published by author. Colombo, Sri Lanka.

Simion, H. G. 1954. Ceylonese beliefs about animals. *Western Folklore* 13: 260–267.

Simion, H. G. & S. A. Wijeyatilleke. 1965. Proverbs from Ceylon about animals. *Western Folklore* 15: 262–281.

Stuart, S. N., Chanson, J. S., Cox, N. A., Young, B., Rodrigues, A., Fischman, D. & R. Walter. 2004. Status and trends of amphibian declines and extinctions worldwide. *Science* 306: 1783–1786.

Stuart, S. N., Hoffmann, M., Chanson, J. S., Cox, N. A., Berridge, R. J., Ramani, P. & B. E. Young. 2008. *Threatened Amphibians of the World*. Edited by Publisher: Lynx Editions, Barcelona, Spain.

Swei, A., Rowley, J. L., Rödder, D., Diesmos, M. L. L., Diesmos, A. C., Briggs, J. C., Brown, R., Cao, T. T., Cheng, T. L., Chong, R. A., Han, B., Hero, J.-M., Hoang, H. D., Kusrini, M. D., Le, D. T. T., Mcguire, J. A., Meegaskumbura, M., -S. Min, M., Mulcahy, D. G., Neang, T., Phimmachak, S., –Q. Rao, D., Reeder, N. M., Schoville S. D., Sivongxay, N., Srei, N., Stöck, M., Stuart, B. L., Torres, L. S., Tran, D. T. A., Tunstall, T. S., Vieites, D. & V. T. Vredenburg. 2011. Is Chytridiomycosis an Emerging Infectious Disease in Asia? PLOS ONE, 6, e23179.

Tennent, E. 1861. *Sketches of the Natural History of Ceylon*. Longman, Green, Longman and Robert's. London, UK.

Van Bocxlaer I., Biju, S. D., Willaert, B., Giri, V., Shouche, Y. S. & F. Bossuyt 2011. Mountain-associated clade endemism in an ancient frog family (Nyctibatrachidae) on the Indian subcontinent. *Molecular Phylogenetics and Evolution* 62: 839–847.

Wickramasinghe, D. D., Oseen, K. L., Kotagama, S. W. & R. J. Wassersug (2004) The terrestrial breeding biology of the Ranid Rock frog *Nannophrys ceylonensis*. *Behaviour* 141: 899-913.

Wickramasinghe, L. J. M., Vidanapathirana, D. R. & N. Wickramasinghe. 2012. Back from the dead: the world's rarest toad *Adenomus kandianus* rediscovered in Sri Lanka. *Zootaxa* 3347: 63–68.

Wickramasinghe, L. J. M., Munindradasa, D. A. I. & P. Fernando. 2012. A new species of *Polypedates* Tschudi (Amphibia, Anura, Rhacophoridae) from Sri Lanka. *Zootaxa* 3498: 63–80.

Wickramasinghe, L. J. M., Vidanapathirana, D. R., Airyarathne, S., Rajeev, G., Chanaka, A., Pastorini, J., Chathuranga, G. & N. Wickramasinghe, 2013. Lost and found: one of the world's most elusive amphibians, *Pseudophilautus stellatus* (Kelaart 1853) rediscovered. *Zootaxa* 3620: 112–128.

Wickramasinghe, L., Vidanapathirana, D. R., Rajeev, M., Ariyarathne, S. C., Chanaka, A., Priyantha, L., Bandara, I. N. & N. Wickramasinghe. 2013. Eight new species of *Pseudophilautus* (Amphibia: Anura: Rhacophoridae) from Sripada World Heritage Site (Peak Wilderness), a local amphibian hotspot in Sri Lanka. *Journal of Threatened Taxa* 5: 3789–3920.

Wickramasinghe, L. J. M., Bandara, I. N., Vidanapathirana, D. R., Tennakoon, K. H., Samarakoon, S. R. & N. Wickramasinghe 2015. *Pseudophilautus dilmah*, a new species of shrub frog (Amphibia: Anura: Rhacophoridae) from a threatened habitat Loolkandura in Sri Lanka. *Journal of Threatened Taxa* 7: 7089–7110.

Wijayathilaka, N., Garg, S., Senevirathne, G., Karunarathna, N., Biju, S. D. & M. Meegaskumbura. 2018. A new species of *Microhyla* (Anura: Microhylidae) from Sri Lanka: an integrative taxonomic approach. *Zootaxa* 4066: 331–342.

Acknowledgements

First and foremost we would like to express our sincere thanks to all herpetologists both in Sri Lanka and overseas for their hard work, which laid the foundation and was the source of information for this book. We are greatly indebted to the following institutions and personnel for their support and insights provided, which went a long way in the compilation of information required to complete this book. The Departments of Wildlife Conservation and Forest Conservation are thanked for the permission given for our studies on amphibians. Former Director General of Archaeology and Mr M. B. Herath (Director, General Services) of the Archaeological Department are thanked for their kind permission to photograph and publish archaeological objects under their care. We are very grateful to Dr Siril Wijesundara, former Director General of the National Botanical Gardens of Sri Lanka, for identifying medicinal plants used in the traditional treatment of poisoning by amphibians.

This book may not have been possible if not for the kind assistance, information and companionship provided in the field and elsewhere by our dearest colleagues, Panduka de Silva, Nayana Dawundasekera, Palitha Chandrarathna, Shantha Karunaratne, Madura de Silva, Ruchira Somaweera, Senani Karunarathna, Suranjan Karunarathna, Parakrama Ekanayake, Imesh Nuwan Bandara, Sanoj Wijayasekara, Suneth Kanishka, Dineth Danushka, Tharaka Priyadarshana and Mendis Wikramasinghe. We are also very grateful to Eranga Geethanjana Perera for the excellent sketches he provided for the book. Wing Commander Rajah Wickramasinghe is acknowledged for the photograph of the silver coin on page 29. We are very much in debt to Sanoj Wijayasekara for his meticulous comments on the *Pseudophilautus* species in this book and for some of the photographs. We are also very grateful to Mendis Wickramasinghe, Usui Toshikazu, Suranjan Karunarathna and Suraj Goonewardena for contributing some of their images to this book.

We thank the IUCN/SSC Amphibian Specialist Group for the Seed Grant to Anslem de Silva and the Mohamed Bin Zayed Species Conservation Fund to Kanishka Ukuwela, which enabled several studies on the amphibians of this magnificent island.

Finally we thank our families for their enormous support for our work and for tolerating our long absences from home during our visits to the great outdoors.

Anslem de Silva, Kanishka Ukuwela and Dilan Chaturanga

Index

Other books about the wildlife of Sri Lanka from John Beaufoy Publishing

See our full range at www.johnbeaufoy.com